Hydrate the Earth

ISBN: 978-2-493291-21-9

© 2021 Éditions La Butineuse
Atelier des Entreprises
Place de l'Europe – Porte Océane 3
56400 Auray – France
www.editions-labutineuse.com

Ananda Fitzsimmons

Hydrate the Earth

The forgotten role of water in the climate crisis

ÉDITIONS la Butineuse

Table of contents

Introduction

The narrative that climate change is caused by the accumulation of greenhouse gases released by human activities, has led to the predominant conclusion that the solution lies in reducing our emissions. While this is true, the deeper that scientists dig into analysing and modeling climate change, the more complex it becomes.

Now, new perspectives are emerging on climate change, focusing on water cycles. It is the main topic I will address in this book, illustrated with concrete and successful examples of water management projects from around the world. Whereas until now the focus has been mainly on carbon cycles, water cycles are now thought to contribute and mitigate climate change just as much.

It should meanwhile be remembered that climate balance cannot be reduced to a single equation. Because of the complexity of ecosystems, it is, in my view, useless to argue whether the true driver of climate change is the carbon cycle or the water cycle: it is both. But the more we understand about how the water cycle affects climate regulation, the more pathways we have to intervene.

Concerning complexity in ecosystems, it should be noted that all living creatures within them play a role, and that all are interconnected through cascades of dependencies. Take for example the wolves in Yellowstone National Park: when reintroduced, they controlled the populations of deer and elk which were overgrazing the vegetation; the vegetation grew back, bringing back small animals, then beavers, rabbits, and some birds. The beavers made dams which generated more wetland and so more aquatic species: water birds, amphibians, then mink and moose. The point here is to show the reality of this interconnectedness, and the complexities how it works in rebalancing an entire ecosystem.

Taking a look back at some key events which have raised awareness of the importance of climate mitigation, I'd like to refer to the *Special Report on Global Warming of 1.5°C*, issued in 2018 by the IPPC (Intergovernmental Panel on Climate Change). This report was highly publicised in the international media and served as a wake-up call, giving us all a dire warning about how we are potentially destroying our planet. Two subsequent reports with further recommendations in 2019 focused respectively on the importance of land use and oceans in saving the planet. Yet we have already been hearing warnings about climate change for decades. The IPCC has been in existence since 1988; in the 2015 COP 21 climate talks in Paris, an idea was tabled by the 4 per 1000 Initiative that we must not only decrease emissions; we could also sequester huge amounts of atmospheric carbon in the Earth's soil. The regenerative movement was born. Such official reports and meetings have built up the momentum, along with many extreme weather catastrophes around the world. They have enabled us to accept the realities of climate change, and realize that we all have a role to play in mitigation.

New words have since crept into everyday parlance, such as the Anthropocene, a geological era which is estimated to have started during the post war period and accelerated with the rapid industrialization and colonization of most the Earth's surface.

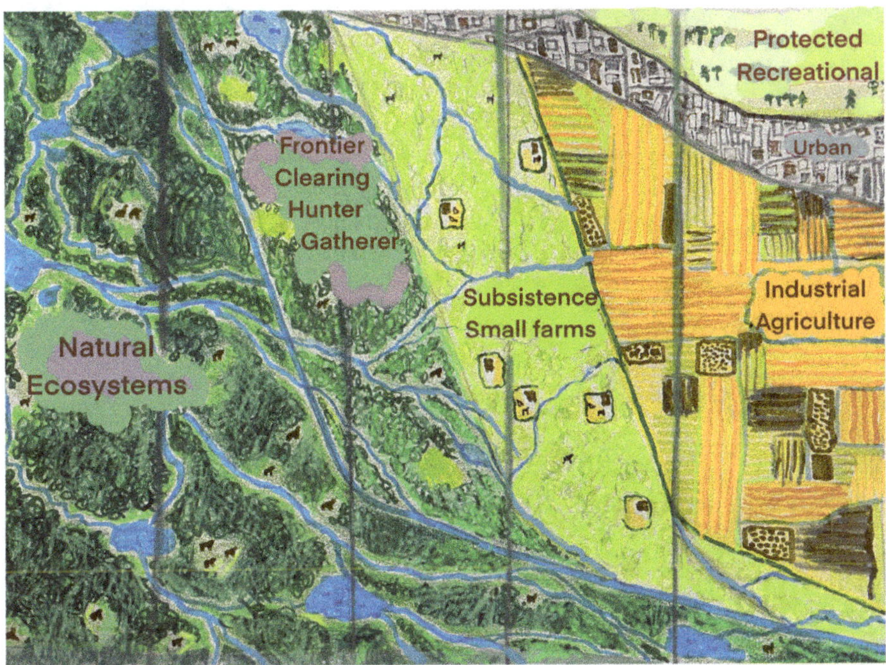

Over the centuries, humans have radically altered the surface of the Earth by removing natural ecosystems and replacing them with human centric systems, which don't deliver the same benefits. In the last panel on the right we see the proportion of natural ecosystems compared to human managed systems since industrial time.

Chapter One
A new water paradigm

Let's dive into these emerging stories about how hydrology is an important driver of our current crisis. One of the dominant voices which coined the term "New Water Paradigm" is that of Michal Kravcik, a Slovak hydrologist and water engineer. In a document by that name, which he published in 2007 along with his team of researchers, they describe two water cycles: the *large water cycle* and the *small water cycle*. The large water cycle refers to the movement of water over the wide landscape. It is also the one studied at school. Water falls from the sky and moves over the land, flowing from higher elevations to lower elevations, gathering in rivers and streams which ultimately flow back to the sea.

The small water cycle refers to the vertical movement of water. Water infiltrates into the earth, hydrating the land. It fills aquifers, which are groundwater storage places. Plants mediate the small water cycle because their roots make the soil permeable, letting water go in deep into the ground. They also transpire water vapour, which cools the earth and the air close to the ground then eventually rises up to become clouds.

Modern industrialized society has disrupted the small water cycle in important ways, and as a result, more and more of the water which falls as rain goes into the large water cycle and is carried back to the sea without cycling many times between the earth and the clouds. Kravcik and his team even posit that this, as much as the melting of glaciers, is contributing to sea level rise.

So how have we disrupted the small water cycle? First of all we have cleared vegetation from the land. Massive deforestation means there are vast areas where ground is not stabilized and protected by trees. Lawns, annual monocrops and

Water runs downwards in rivers to the sea, but it also seeps into the earth and is breathed out by vegetation.

shrubs don't have such deep roots and don't transpire as much as trees. Asphalt and concrete don't transpire at all and don't absorb rainfall. In cities and industrial areas sewage pipes and drainage ditches carry the rainfall away, straight back to rivers, often with a nice load of pollution and back to the large water cycle.

In agriculture, bare tilled soil loses not only its carbon but allows water to evaporate quickly as well. Heavy machinery has compacted agricultural soils, so that water runs off the surface rather than seeping in. Excess water is channelled into drainage ditches or drainage tiles and quickly evacuated. Agricultural drainage practices are concerned with preventing standing water and floods; they are less concerned with good water infiltration, so again, we are feeding the large water cycle but not the small one.

The planet is desertifying

There is such a thing as a natural desert, a dryland ecosystem where the vegetation is adapted to a dry climate, but desertification is another thing. Desertification is caused by the degradation of land, due to unsustainable land management. The same practices that are disrupting the small water cycle: deforestation, overgrazing, tillage, monocultures, urban sprawl, are creating larger areas of desertification around the planet.

There are vast deserts on the planet which were once fertile. The Fertile Crescent, an area in the Middle East spanning what is now Syria, Lebanon, Palestine, Israel, Jordan and Egypt, is called the cradle of civilization. Once a lush and fertile land, it was the birthplace of modern agriculture. There originated the practices of plowing, irrigation, raising domestic animals, and the culture of plants such as wheat, barley, chickpea and lentil. But today, no agriculture in that region is possible without extensive irrigation. The once fertile soils are contaminated by salinization, which is what happens when groundwater is used for irrigation over long periods of time. Minerals from the earth and synthetic agricultural amendments such as fertilizers concentrate in the ground water and accumulate in the soil, eventually contaminating the soil until it can't sustain life.

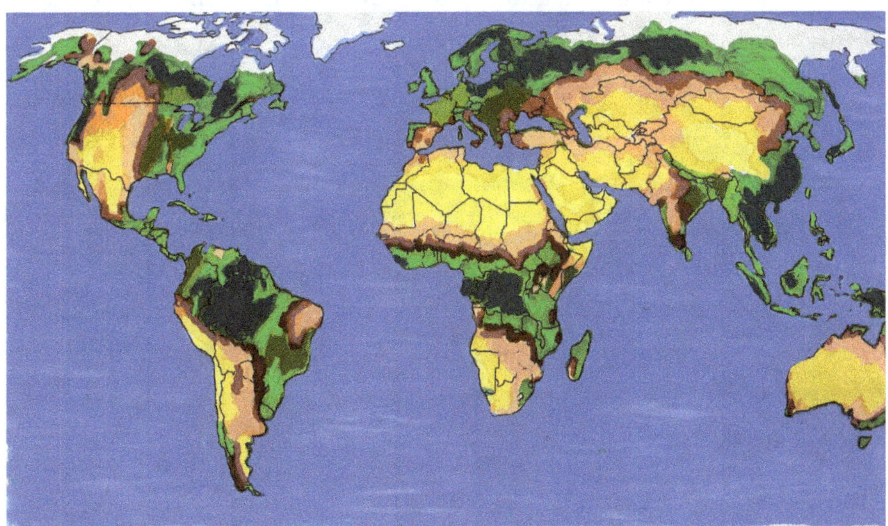

Once fertile areas are becoming increasingly arid, due to poor land management and deforestation.

The same thing is happening in other places. In the southern USA, notably California and Arizona, where most of North America's fresh produce is grown, little rain falls and agricultural fields are kept in production by irrigation, pulling the water from major rivers and underground reserves. But these reserves do not receive enough rainfall to refill them, so over time they are being depleted. When you travel, outside of the agricultural fields, you see dried up riverbeds and desert landscapes. And yet supermarkets over North America are selling lettuce produced in deserts with irrigation that is coming from diminishing aquifers.

Where I live in Eastern Canada, many people do not take seriously the possibility of a water crisis, mostly because we have plenty of water here. And it is true: often we have too much. When the snow melts in the spring, we frequently get flooding, and with climate change, that is happening more often. We have serious flooding events where people have to evacuate their homes, and cornfields can't be planted. But in 2020, even after spring flooding, we had a six-week drought in the summer. Farmers who grow broad field crops are not equipped for irrigation because they never needed it here. That summer they became worried, and now wonder if, after all, they will need to install irrigation systems in the fields. Fruit and vegetable farmers are already doing it.

Too much rain and not enough are simply two sides of the same coin. Monsoon rains are a part of desertification. Even extremely dry places get torrential rains sometimes. When it rains, it rains more, but then it does not rain for a long time. There is no place on the planet which is not touched by the disruption of water cycles. When we think of deserts, we may think they are far away and don't concern us. But we are doing the same things everywhere. We clear vegetation and mismanage water which eventually leads to soil degradation and desertification. It is only a matter of time unless we begin to do things differently.

How does the water cycle regulate climate?

Walter Jehne is an internationally acclaimed climate scientist and microbiologist, and is the founder of Healthy Soils Australia. He is another of the passionate voices saying that restoring the water cycles is our best chance to avoid catastrophic climate change. According to him, the rise in atmospheric carbon is a symptom of climate change, and not the biggest cause. Because water vapour is also a greenhouse gas, responsible for 95% of the heating and cooling mechanisms on planet Earth, it plays an even more significant role than CO_2.

He goes on to describe those mechanisms and I will try to summarize what he says because it sheds light on how we need to change the way we manage both land and water.

Albedo effect

The sun radiates 340 watts of heat per square meter of earth surface every day. Some of it is absorbed while some of it is refracted off the surface. In order to keep the balance of warming and cooling of the Earth's temperature, part of the heat that comes into the atmosphere each day has to escape back into space. This process is slowed down by the Earth's atmosphere, which traps and slows the process of radiation going back into space. We call this the greenhouse effect and it is necessary to buffer this heating and cooling process so that we can keep a comfortable temperature on the Earth's surface to sustain life as we know it. There is an input and an output rate, which need to be in balance.

Two things are happening with climate change which are altering this balance. One, well described and understood by most people now, is the accumulation of more greenhouse gases in the atmosphere. Greenhouse gases trap more heat close to Earth and slow its escape back to space.

The second factor is less understood. It has to do with how heat is absorbed or reflected as it strikes the Earth's surface. If solar radiation hits a dark surface,

Solar radiation penetrates into dark surfaces causing heat islands but bounces off of clouds and vegetation.

more of it is absorbed. If it hits a light or reflective surface, it bounces off and is refracted back into the atmosphere. So if the sunshine hits an asphalt surface, or bare soil, or concrete it will be absorbed. But if it hits a surface covered with plants, or a cloud, or ice, more of the radiation will bounce off and it will not warm the lower atmosphere.

This principle is called the albedo effect and it accounts for how we create urban heat islands and why it is so much more comfortable in the shade on a hot day. But Walter Jehne throws in that the difference between an absorbent dark surface and a reflective surface is exponential. When radiation hits a dark surface the temperature of the radiation that hits it is multiplied to the power of 4 when it is released. So that is Temperature x Temperature x Temperature x Temperature that is eventually radiated back out of that surface. This means that the difference in heat accumulation between a plant-covered ecosystem, or an ice-covered landscape versus a city surface or a bare plowed field is enormous. Think about how much of Earth's surface is now covered by cities, industrial sites and plowed agricultural fields compared to how it was a couple of hundred years ago.

Transpiration

Besides creating shade and refracting light away from the Earth's surface, plants have an even more important role to play in cooling our planet. Plants transpire. This is much like how humans cool themselves by perspiring. When we sweat, the water of our sweat evaporates and that carries away excess heat. In the same way, when water evaporates from any surface, it cools that surface.

When plants transpire, they actually use heat energy to convert water into water vapour. A single mature tree can circulate hundreds of litres of water in one day. The roots of plants pull water up from the depth of the Earth, convert it to water vapour and exude it into the atmosphere through their leaves. This is an important way that Nature has devised to circulate water from deep in the Earth back up to the higher atmosphere. They absorb heat and release water vapour. Of course this solution depends on there being water in the ground and a diversity of plants with roots at different depths!

The transpiration of plants means the release of water vapour into the air, and it is this water vapour that coalesces to form clouds. Clouds are important, not only because they make rain, but also because they deflect heat back into space. Normally 50% of Earth's surface is shielded from the sun by cloud cover. The radiation that strikes it never gets to the Earth's surface and this contributes to temperature moderation. Of course, if there is less vegetation transpiring, less water vapour is in the air and less clouds can form.

Trees are critical in both the carbon cycle and the water cycle. They pull carbon from the air, convert it to liquid and put it into the ground through their roots. They also pull water up from the ground, convert it into vapour and transpire it back into the air, cooling the ground, forming clouds and stimulating rainfall.

What makes it rain?

Water vapour and clouds are needed to make rain. But it's more complex than that. In order for clouds to begin to rain, it takes what is called precipitation nuclei to 'seed' them and form rain droplets heavy enough to fall. There are three things which can serve as precipitation nuclei to 'seed' the clouds. One is ice crystals. Another is salt, which is why it rains more over the oceans. The third option and the most important over land far away from the coasts, is a kind of bacteria which is found in the water vapour of tree transpiration. Areas which are covered in forest get more precipitation. The transpiration of trees is critical to regulating rainfall inland.

It is not uncommon in tropical rainforests to get daily rainfalls. Because the tree roots go deep, rainfall can penetrate into the earth. Water is stored right to the depth of the roots. Other plants keep the humidity from evaporating and moisten the air. In a rainforest, one droplet of water may circulate between the Earth and the atmosphere many times in a very rapid cycle. Every time it cycles it brings cooling, both on its way up and again on its way down.

Haze

Aside from deforestation and removing plant cover, human civilization has created another problem which keeps rain from falling: persistent hazes. Haze is caused by particulate matter binding to tiny water droplets. Between pollution from industrial processes, fossil fuels, forest fires, vehicles, dust from eroded topsoil, the Earth's atmosphere contains far more particulate matter than ever before. The droplets bound to these tiny dust particles are very small and are called microdroplets; it takes 1 million microdroplets to make one raindrop. They are so small that they float in the air and don't coalesce into raindrops. They just hang in the air and actually absorb radiation, so they keep excess heat close to the Earth's surface.

This is one reason why we now get "stuck" weather patterns, when hot hazy air hovers in an area for many days. A good downpour of rain would clear the air, push the particulate matter down to the Earth and allow the air to clear and the heat to escape to the upper atmosphere. But if the water vapour is trapped in the haze particles, the hot air mass doesn't move.

The Biotic Pump

It is well known that it rains more on forest and coastal areas than on cleared inland areas. But how does this happen? How does the wind that moves the clouds work?

Wind is a movement of air that flows naturally from areas of high pressure – consisting of warm air – to areas of low pressure – consisting of cold air. Low pressure zones are therefore necessary on continents, so that the wind can push inland the clouds that have formed over the ocean.

Let's now turn to condensation. When it occurs, water changes from a gaseous state (as water vapour) to a liquid state (as water drops). The immediate result of this transformation is that the pressure of the air is lowered, and it is no longer charged with water particles. If a continuous condensation process is maintained on the continents, they become a permanent low-pressure area.

Mature forests, with their multiple layers of canopy, from undergrowth to slow-growing trees and shrubs, to large trees and vines, provide an incredible evapotranspiration surface. The air in a dense forest contains more moisture than the air above the ocean. When plants convert large amounts of water into vapour, the air moves upwards and condensation takes place.

Through this phenomenon, inland forests, by generating low-pressure areas, play a major role in drawing water from coastal areas. Dense mature forests constitute a huge low-pressure system that draws clouds inland.

This biotic pump theory, which gives a major role to the phenomenon of condensation in the circulation of air and water on Earth, is still not very well known and controversial among climate scientists. One of its major proponents is Anastasia Makarieva, a Russian physicist, who is a fierce defender of the world's largest remaining forests. She maintains that the few

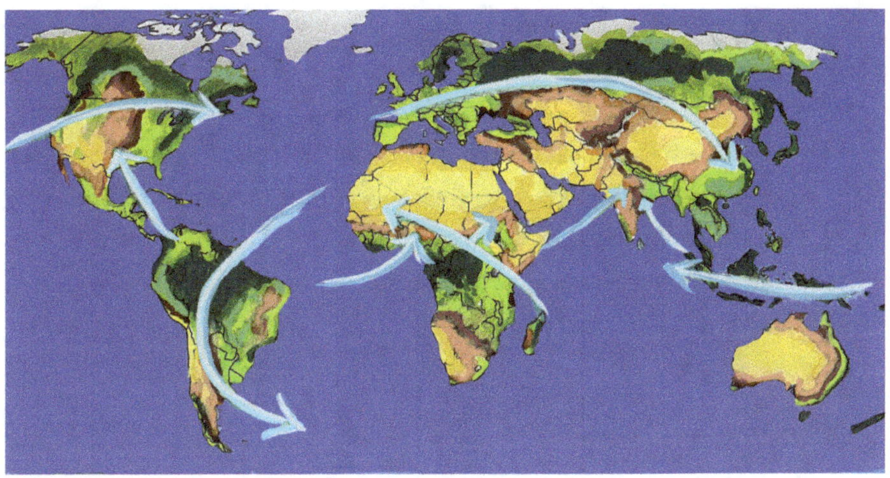

"Flying rivers," currents of moist air generated by forests bring humid air across the continents.

remaining extensive forest areas, such as the Amazonian rainforest and the boreal forests of Russia and Scandinavia, are not only responsible for local humidity, but also allow moist air to circulate for hundreds of kilometres thanks to wind currents created by the biotic pump effect. The jet stream carrying the moisture generated by forest transpiration has been dubbed the "flying river". Most of the rainfall in South America is thought to be generated by air currents from the Amazon. In China, most of the precipitation is thought to be generated by boreal forests in Russia/Scandinavia.

The biotic pump theory, developed by Makarieva with her mentor, Victor Gorshkov, in a paper entitled Hydrology and Earth System Sciences in 2007, contests the conventional understanding of how meteorologists believe air currents move. Conventional wisdom holds that clouds generally form over the ocean, to be driven inland by winds caused by temperature differences of the atmospheric masses alone. Yet forested areas inland such as the Amazon often receive more rain than coastal areas. This would not be true if the coastal winds were the only source of the water. The idea that condensation generated by forests is actually responsible for creating winds of moisture is radical, and goes a long way to explain how moist air is generated inland.

If Makarieva is correct, the importance of protecting what remains of the world's major forest systems cannot be overstated. Makarieva and Gorshkov are also adamant that mature, natural forests will most effectively maintain the water cycles, not artificial plantations and certainly not tree monocultures. A mature forest ecosystem has evolved over thousands of years creating habitat for a complex balance of species, which regulate the moisture flows by working together. A younger plantation of trees on previously cleared land will not have the same impact.

Chapter Two
Regenerating the water cycle

Since human interventions have disrupted the water cycles, it is also within our power to recognize the impacts of our actions, and make changes to reverse the harm by regenerating water cycles. There are two fundamental and complementary ways to restore the functioning of the small water cycle. Remember that the small water cycle is the vertical movement of water that sinks into the ground and returns to the atmosphere through evaporation and the transpiration of plants. If water doesn't sink into the ground to adequately hydrate the earth, it can't be returned to the air, cooling the local temperatures. Instead, it runs off the land quickly, taking particulate matter with it, and pollutes the rivers, streams and lakes. Ultimately polluted water runs into the ocean, contaminating it as well. The water flowing quickly over land picks up organic matter and contaminants such as fertilizers and other chemicals from industry and deposits them into the water bodies.

The disrupted small water cycle creates a negative feedback loop. The less vegetation there is, the more quickly water evaporates. The more quickly water evaporates, less water is stored in the earth. Drier conditions lead to less vegetation thriving.

So the two fundamental actions we can take to remedy this negative loop are to:

• allow available water to sink deeply into the ground by restoring the quality of the soil;

• make sure that the landscape captures the water that falls so that water stays longer in a local area.

So let's take a dive into the how of each of these, beginning with soil regeneration.

The soil carbon sponge

Walter Jehne's preferred approach to restoration of the water cycle is the regeneration of the soil carbon sponge. Indeed, this approach is the first line of defense, the cheapest and most natural solution, and the one to be tried before anything else. Many people, in trying to explain the soil carbon sponge, use a simple little demonstration with two plates, some flour and a piece of bread.

One plate has loose flour on it. When we sprinkle water over the flour, it runs off in rivulets. The flour is dry and hydrophobic, so water doesn't sink in. This is what happens with very parched, bare earth. It repels the first rain. When you pour water over the second plate with the piece of bread, it soaks right in and doesn't run into the plate until it's completely saturated. This represents rainfall on healthy soil. It acts like a sponge. Why? Because it contains microorganisms and lots of organic matter which give the soil structure. Healthy soil has aggregates, meaning it sticks together in little clumps made up of mineral matter, organic matter and microorganisms, glued together by glomalin, a glue created by microorganisms. This structure leaves spaces which contain air and water within the soil. The soil is porous and acts like a sponge.

The USDA NRCS, the United States soil conservation agency, estimates that every 1% increase in SOM (soil organic matter) gives soil an increased water retention capacity of 20,000 gallons per acre. This is very significant when it comes to surviving a drought or a flood. When there is too much rain, a field rich in organic matter can absorb it for use by plants later on. When there is a prolonged period with no rain, a soil that stores significant amounts of water can keep plants alive without irrigation for quite some time.

It is not generally known that soil needs plants growing in it to keep it alive and healthy. Plants breathe in carbon from the air and convert it through photosynthesis to liquid carbon, which they exude into the soil through their roots. This liquid carbon feeds the soil microorganisms. In this elegant symbiosis between plants and soil microorganisms, the latter in return deliver mineral nutrients from the soil in a form which plants can assimilate. This exchange of carbon for minerals is the very basis of soil fertility. This is how Nature has allowed terrestrial ecosystems to thrive for millenia without men adding fertilizers.

Modern industrial agriculture evolved in ignorance of this natural soil ecosystem. Many of its methods destroy the soil carbon sponge. Inputs such as salt based, synthetic fertilizers and poisonous chemicals used to control pests and weeds, kill off many of the organisms which are basic to healthy soil function. Fungus is particularly sensitive to these chemicals and to tillage and soil compaction. It is a general belief that fungus is a pathogen, and indeed, there are many kinds that are. But there are also a host of beneficial fungi which work alongside beneficial soil bacteria to form the basis of a healthy soil ecosystem. Fungi create vast webs of tiny filaments which transport nutrients and create a communication system in the soil between plant roots and bacteria. Well known mycologist and lover of fungus, Paul Stamets called this below ground fungal network "the wood wide web". When we destroy this essential fungal network with chemicals and excess tillage, it leaves the field open for pathogenic fungus to take over, creating another negative feedback loop.

The regenerative agriculture movement, which has been gaining momentum around the world over the last number of years, is an effort to shift agriculture towards practices which can produce enough food to feed the population and at the same time restore the healthy soil ecosystem. A lot of this momentum comes from the climate change conversation. People around the world are coming to understand that drawing carbon out of the atmosphere through good soil health practices can sequester carbon in the ground and help to offset emissions of human activities. However, the role of water in all this is seldom talked about, although fortunately all the regenerative soil health practices benefit the water cycle just as much as the carbon cycle. That saves us from having to argue about what to do. Yet we still need to understand and document the benefits to the water cycle in addition to measuring and rewarding carbon sequestration. Then perhaps we can broaden the conversation from an exclusive focus on storing carbon to a more holistic view of all the ecosystem services which are so important to mitigating climate change. So, what are these soil regeneration practices?

Regenerative land management practices

Regenerative land management practices are based on a series of principles for optimal soil health and help to counter the negative impacts caused by destructive modern methods. There is no one method which can be called regenerative because the context of each piece of land and each production system varies tremendously. Once land stewards understand

the principles of regenerative agriculture, they can choose to integrate the methods which best suit their situation.

Minimize soil disturbance

Plowing and turning over the soil has been the cornerstone of agriculture since it first began. And yet Nature never leaves soil bare. The only time you will see bare soil is if it has been disrupted by some force such as humans or animals digging it up. When the surface of the soil is disturbed, it is like getting a cut or wound to our body. What holds everything together has been opened up, leaving what should be protected inside exposed. In Nature, as soon as the cover of the ground has been disrupted, weeds start to fill in the void. We may think bare soil with neat rows of crops is tidy, but Nature will always be trying her best to get some weeds growing to fill it back in. The history of agriculture could be seen as the human fight with Nature to control weeds. In modern times the invention of herbicides has taken over where tilling was the only way to remove unwanted vegetation.

The reason Nature is so insistent on filling in bare spaces with plant growth is that bare soil leaks both carbon and water into the atmosphere. Organic carbon in the soil volatilizes when it is exposed. Moisture evaporates when exposed to air and sunshine. When the fine networks of beneficial fungus – are cut up regularly with the blades of a tiller, they have a hard time growing back and maintaining a healthy population. The beneficial bacteria in the soil need shelter and moisture. They do not take kindly to being baked and having their homes turned upside down, any more than we would welcome having our cities shaken up and knocked over by an earthquake.

So the first principle of managing land to promote healthy soil is to find a way to disturb it as little as possible and to try to imitate the way Nature builds soil. This is not easy when we fundamentally want to manage what grows for our own benefit, but regenerative land management principles show us ways that we can minimize the harm we do to the soil ecosystem, conserving the carbon and the water necessary to keep it fertile.

Soil Cover

The second principle, keeping the soil covered, goes hand in hand with minimizing disturbance. If we must disturb the soil surface and remove the vegetation that is growing there, at least we can do that for the shortest time possible and replace it with something else that will keep the underground life intact. This can be done with special synthetic cloth tarps or with dry organic matter (called mulch) but the best practice is cover cropping.

Cover crops are live plants that are grown, not for our benefit, but for the benefit of the soil. They can be planted after a crop is harvested, to keep the soil fed and covered over the fall and winter, or early in the spring before something else is grown. They can also be planted in between the rows of cash crops, by choosing a cover plant which will be sown in such a way as to not interfere with the main crop. For instance, clover which is low growing can be planted between rows of a taller crop. There is a whole emerging science that studies the benefits of cover crops for commercial plants, as they can enhance their growth by providing nutrients or protect them from pests. When done well, there are many benefits to this practice. They help farmers to minimize their inputs, reduce their use of fossil fuel and grow healthier crops at lower cost.

The no-till movement began in the 1980's with direct seeding. This involved leaving the rubble of the previous crop on the soil surface and then planting the next crop by using a special planting drill that pushed the seed into the ground through the dead organic matter. Direct seeding did help a lot with the problem of soil erosion, but continued studies revealed that it did not do as much to sequester carbon as expected. Now the no-till movement is focussing more on the use of living cover crops to cover the land for as long as possible in the yearly cycle. We'll talk more about why in the next chapter on living roots.

Perennial Plants, Living Roots

Modern commodity crops grown around the world are generally annual crops which need to be sown again every year. Annuals have shallow roots because they don't have time to develop deep roots. In terms of ecosystem services, such as carbon sequestration and water infiltration, roots have a very important role to play. Deep rooted ecosystems such as forests and grasslands sequester more carbon and more water than annual cropping systems. When more and more of these natural ecosystems are destroyed and converted to annual cropping systems, the capacity of the ecosystem services is greatly reduced.

Modern agricultural plants, such as soybeans, have been developed by humans to maximize the aboveground growth, to get more harvestable beans. But in our quest to maximize food, we sacrificed food for the soil microbiome and disturbed the carbon and water cycles. We have taken away the organic matter as harvest and not kept enough of it to return to the soil ecosystem. So can we find a better balance by integrating more perennial crops into our agroecosystems?

Planting rows of tree crops between annual crops, planting edible forest gardens and integrating more perennial food crops into our diets are some practices which regenerative land stewards have been using in order to get to a better balance between human needs and Nature's needs.

All plants play a role in mediating the carbon and water cycles, but those with deeper root systems are much more efficient. Trees have the deepest roots, so they have the biggest root surface area. Perennial grasses are also important since their root systems can become very deep and thick. Roots feed the microorganisms in the soil with liquid carbon; the plant makes it by using the sun's energy to convert the carbon it breathes in from the air. This is why having the ground covered with living plants for as much of the year as possible is more effective than just covering the ground with dead organic matter. Dead organic matter contributes to soil carbon when it breaks down, but studies have shown that the carbon from living roots has a much larger contribution to effectively building soil.

In a demonstration that shows the effectiveness of having living roots in the soil for preventing soil erosion and protecting waterways, we take three samples of field soil. One is bare soil, one has a covering of dead plant debris and the third has living plants growing in it. When we pour water onto the three patches of earth and collect the water that comes out into a container underneath, we see under the bare soil much of the water passes right through, depositing a great deal of muddy sediment. Under the patch that is protected with crop residues, less water comes out the bottom with less sediment and the water is somewhat clearer. But in the patch with living plants, most of the water is held in the soil and the small amount of water that comes out the bottom is clear

Biodiversity

In any natural ecosystem, there is a tremendous amount of biodiversity. Humans tend to create ecosystems with minimal diversity. Think about how the farmland you drive by looks with its vast monocultures. Or how the average landscape around a residential home looks with a few shrubs and flowerbeds and a mowed lawn. Compare this to one square meter of any wild landscape. You will be able to notice dozens of species in the smallest square of a natural landscape.

The study of ecology is discovering that each species in a natural ecosystem has a function which contributes to the resiliency and sustainability of that system. The more we remove species the more vulnerable the ecosystem becomes. This is evidenced by the fragility of our modern agroecosystems. They are plagued by problems and diseases which need to be managed with chemical inputs.

We are learning that plants exude a multitude of biochemicals which have an impact on their neighbouring plants. They grow better in a community of compatible species than they do in a monoculture. Microorganisms also give off biochemicals which stimulate the functions of other microorganisms and plants. Science is just beginning to understand the amazing complexity of these interactions and to study the interrelationships between microorganisms, plants and animals.

We have learned that the human body contains only 10% human cells and 90% microorganisms which have evolved alongside us and without whom we could not survive. Our microbial companions are responsible for making sure we can digest our food and resist disease. All of Nature is made up of microbial communities. The soil contains billions of microorganisms and is to plants what our gut microbiome is to us. Soil microbes are essential to a plant's ability to assimilate nutrients and resist disease. When we unknowingly destroy the diversity of microbes in our gut or in our soil, we end up with a weaker system which has a diminished capacity to maintain healthy functions.

Regenerative land management aims to increase the diversity of plants, microorganisms, insects, birds and animals in the managed ecosystem to imitate the natural system. As we go along, we learn more and more about the synergies that are so important. Biodiversity isn't just about throwing together as many species as possible. Evolution has developed synergies that work. We just don't understand them all. For example, in the research on cover crops, we are learning which plants to put together to achieve the optimal benefits.

In Nature, certain plants grow together with certain mushrooms, insects, birds. They have evolved together to create synergies which help all species live together. This knowledge needs to be understood and applied. Yet we will never understand all of it because it is so complex. The more we observe Nature, the more clues we will get to understand why some things work better together. But in the absence of being able to understand all the complexity, regenerative systems strive to include as much biodiversity as possible.

On a regenerative farm, we add more plants into the rotations. Sometimes we may combine many plant species at the same time in a field. Each plant exudes carbon through its roots as well as biochemical compounds unique to itself. The signals coming from the plant determine which microbes will thrive around its roots. The more diverse plant species we have, the more microbial diversity will be in the soil. Different plants will attract different insects, birds, fungal species.

We may also have areas left more or less wild, for example around water or strips left as habitat for pollinator insects and birds. These patches of diverse habitat provide space for wild species to coexist with the more tended areas. Regenerative farmers notice the return of wild species of all kinds to the farm as they increase their biodiversity and the whole area begins to function more like a natural ecosystem.

Organic Inputs

Before the invention of chemical fertilizer, herbicide and pesticides, Nature was able to grow lush healthy plants. This is because the soil ecosystem evolved to cycle nutrients in a closed loop system. All the nutrients which are the building blocks of life get recycled back into the earth to keep the system going perpetually. This is the essence of true sustainability. The waste products of each plant and animal decompose to release nutrients back to the system. Microbes are specialized to break down every form of organic matter once it dies.

We have disrupted this natural recycling system in various ways. As I have already mentioned, by tilling the soil and reducing the biodiversity of species, we have limited the microbial populations in cultivated soil. But we also remove most of what has been grown in an agricultural field. **We harvest the plant's biomass and ship it off to where people consume the food. The waste matter never returns to complete the cycle. Our food waste rarely ends up back in the field.** Our faecal matter is flushed into water treatment systems and doesn't return to feed the earth. So the cycle is broken and we need to replace what we have taken from the soil with fertilizer.

When we try to regenerate the function of the soil ecosystem, we have to put back what was broken by letting go of the chemical inputs and introducing organic inputs that replicate a natural system. This does not mean replacing synthetic fertilizers with certified organic ones! It is not essential to be fully organic to regenerate the soil ecosystem either. We can reduce the chemical inputs gradually while transitioning and use them judiciously and selectively while building up the soil life.

It does mean that we have to feed the soil with a large quantity of organic biomass to replace what is taken away. A part of the cash crop is harvested and leaves the farm, but the residues of the crop stay on the soil. These plants are killed off but their biomass remains as residue, covering the soil, feeding soil microbes and are gradually converted to soil organic matter. Regenerative farmers do this mainly with cover crops which are planted just to feed the soil.

It is possible to feed the soil with organic nutrients in other ways, but the use of diverse plant cover is by far the most effective. Using compost, mulches of dried organic matter, organic fertilizers and biostimulants helps to kickstart the healthy function of the soil ecosystem, but above all, the soil needs to be regularly fed with the biomass of diverse plants. We also need to transition away from any chemicals which kill off our valuable microbial allies.

Integrate animals

Animals of all sorts are always part of a natural ecosystem and they have an important role to play. Industrial agriculture has separated the animals from the plants into specialized systems, to the detriment of both. Sure, it is easier to specialize in only one thing but in the long run, we have again unwittingly broken a cycle which turns out to be vital.

Plants stay in one place, while animals move, making them important companions for plants. They deposit their wastes on the ground to feed the soil and they move seeds and pollen from one place to another. When animals nibble on plants they create reactions which help them to grow better.

Allan Savory is an ecologist who was working on the problem of desertification in Africa in the 1950's and 60's. He believed that overgrazing was the cause of desertification. Based on his recommendations which were backed by a committee of scientists who shared the belief that the trampling and grazing of too many animals was the cause of land degradation and desertification, 40,000 elephants were culled in Rhodesia until 1969. The culling of the elephant herds did nothing to reverse desertification. In fact it only became worse.

Allan Savory went on to develop the theory of "holistic grazing" and he founded the Savory Institute. This organization now works around the world to raise awareness of this method of managing livestock, inspired from wild herds, to regenerate grasslands. The idea is that in climates where the natural ecology is grasslands, the ecosystems evolve with massive herds of grazing animals, such as elephants, bison, goats, deer and horses. The animals are followed by predators like wild cats or wolves. In order to protect itself, the herd moves closely together and constantly travels over the landscape, eating, trampling and defecating. When the herd stays in one place for too long, we have the problem of overgrazing. But when the animals move and only return to the same spot once the grasses have had a chance to recover, the animals' presence actually helps the grassland to thrive.

In a landscape which only gets seasonal rains, if there are no animals the grass grows up to seed and dries. The dry grass oxidizes instead of decomposing, meaning it suffocates new grass from coming up and eventually causes bare patches of soil, which get taken over by woody bushes and scrub. In these areas, carbon and water are released from the soil. Grazing animals browse the grass, causing it to regrow more vigorously and to send out more shoots. Their trampling helps dead grass to easily decompose while their manure and urine fertilize the soil. As long as the herd moves continuously, the result is a healthy grassland with deep roots and a vigorous covering of grass with no bare patches.

In numerous case studies in many different countries, where holistic grazing has been implemented, a marked difference to the regional water cycle has been observed. Simply having the land surface covered with deep rooted healthy grasses has meant that the rain accumulates in rivers and water holes and stays there instead of evaporating or running off immediately. With water retained in the landscape, fertility returns.

As wild animal populations disappear, domestic animal herds are raised in tight confinement on feedlots, and grasslands are converted to fields of annual monocrops. We are losing a critical habitat which has historically sequestered large amounts of carbon and held water in the soil. These landscapes are the ones which are desertifying the fastest and they cover a huge part of the earth's surface. The impact of losing grasslands on such a large area is a major contributor to the disruptions of the carbon cycle, the water cycle and ultimately to climate change.

The principle of integrating animals to regenerate the land includes bringing grazing animals back into the landscape, especially in areas where the natural ecosystem was grassland. Even in wetter climates, it is beneficial to have some animal movement on the land. Forested areas once had deer and other wild animals moving through them, leaving their waste material, stimulating growth by nibbling and helping to avoid overgrowth. Integrating animal rotations in farm systems everywhere adds biodiversity and provides an important source of organic material to feed the soil.

Of course, the integration of animals includes smaller creatures as well. Making sure there are birds and pollinator insects is a natural way of creating a balanced ecosystem, and the presence of wild habitat and creatures is important for controlling pests and diseases without chemicals.

Chapter Three
Ecosystem preservation and restoration

In addition to regenerating land to create the soil carbon sponge, we also need to look at water dynamics from a landscape level perspective. Water naturally collects in certain places because of the shape of the land. The nature of water is to flow downhill. Any place where water falls, it will flow down the slope until it hits a bowl-shaped area where the sides are higher than where the water sits. There it will stay until it is either absorbed into the earth or evaporated into the air. These places where water sits for a longer time are very important for slowing down the movement of flowing water back to the sea. They keep the balance between the large water cycle – horizontal movement – and the small water cycle – vertical movement.

Water that stays on the land for a longer time is critical for several reasons. It allows water to percolate slowly into the land, keeping it hydrated and feeding the reserves of underground water. It also allows evaporation which humidifies the air. If there are plants in these wet areas, the process of transpiration keeps the atmosphere cooled and liveable.

In our industrialized society, we have long considered these places a waste of space because we can't use them for anything we value. We can't build or grow crops on them so we have tended to fill them in in order to turn them into something which makes money. We drain away the water and bring in fill to make them "productive". How short-sighted humans have been! Only now are we beginning to understand that these wet areas provide us with something critical to our existence. They are necessary to maintain the carbon cycle and the small water cycle, which make our planet amenable to life by regulating the climate.

We also have to mention the value of biodiversity in ecosystems which store and collect water. In nature, the places where two ecosystems meet is where we find the most biodiversity. Where the forest meets the field, where the land meets the water, we find an incredible diversity of species that are not found anywhere else. Why is this so important? Remember the discussion about the wolves in Yellowstone? Every species has an ecological function and Nature has a built-in resilience, by having many different species that do something similar in a slightly different way. If there is a problem with one species not being able to do the job, there can be several others who make up for the lack. Even though we may not understand what the importance may be of a certain bird, insect or plant which lives in a particular habitat, there may be other species who depend on it to survive. If we destroy a wetland ecosystem, there will be multiple species requiring this type of habitat, who will all be wiped out at the same time. How long will it take us to notice that we are losing an ecosystem function that we took for granted and which was actually important for our well-being on this earth?

Let's take a closer look at some of these water preserving ecosystems that we have undervalued.

The ecological importance of wetlands

At least half of the natural wetland areas in industrialized countries have been destroyed and only recently has there been an interest in preserving and restoring them because we have started to understand why they have value.

Natural living systems have an ebb and flow to them like a heartbeat: wet and dry season, cold and warm season, productive and dormant season. They are self-regulating. They have built-in balancing mechanisms. Wetlands are a part of this self-regulation of the water cycles. In the season where there is a lot of rain in a short period of time, water collects in the wetlands

and slowly dissipates through the dry season. Without wetlands there are floods, followed by droughts. Wetlands give the excess water which can't be absorbed by fields and concrete structures a place to go. That water will sit and be gradually absorbed during drier periods. Where there is water stored in the earth, this process is gentler and kinder. The streams may dry up in the summer, the water level of ponds and lakes may go down, but if the earth is holding water, if the groundwater reserves have been replenished, there is resilience. The vegetation can endure the dry, hot times for longer. The atmosphere is more comfortable for the people and the animals.

Wetlands also serve as collectors of sediment and contaminants. Matter that is eroded from the surface of the land settles in the lowland collector areas of wetlands. If not for wetlands, this matter would keep flowing in fast moving waters and contaminate the lakes, rivers and oceans. Wetlands purify the water because they slow the flow, allowing sediments to be deposited. The water sinks slowly into the ground. Contaminants are filtered out by the earth, so that when the water returns back to the water table or the stream, it is cleaned and purified.

When heavy rains fall, a lot of organic matter is picked up by the rivulets of water that wash over the earth's surface. Regenerative land management plays an important role here by applying methods to minimize the erosion of wind and rain. However, inevitably – as when we wash an object by rinsing it under a jet of water, the force of a hard rain will dislodge debris and carry it downhill. The faster the flow of water, the farther the particles of matter will travel. Any hollow in the land, any place where water is caught for a time without continuing its journey downhill, helps the sediment to settle.

Many of our water purification systems work on this principle. For example, a septic system catches the water that is flushed down a toilet or goes down the drain in a house. It is caught in a holding tank where the flow slows down. The solids settle to the bottom of the tank and the clear water flows through an opening at the top. The solids then decompose with the help of microbes which occur naturally in the sediment and the clarified water continues its course towards the lowest point, eventually a lake or ocean.

The sediment may be topsoil, organic matter, something which is valuable to build soil fertility. When it is deposited, it can be transformed by soil microbes into something which feeds life. But when it is washed away into the water system, it becomes a contaminant. Excess organic matter in waterways causes another kind of microbial growth which is less interesting for the forms of life that suit us humans. It clogs up pipes, fills in rivers and makes invasive algae and plants grow in ponds and lakes. When it is deposited on land it creates fertility for plants. Excess phosphorus and nitrogen, when they end up in waterways become contaminants, but when they are left on the land, they create fertility.

Even contaminants, such as pesticides and industrial wastes created by humans, do much greater damage when they are transported into waterways. Many contaminants can eventually be decomposed if left to compost for a long time. When they are transported and diluted in water they travel much farther and do more damage than they would otherwise.

Protecting the blood of the Earth

Rivers and streams are the blood and veins of the Earth. They transport water and visit many corners of the land before they empty back to the sea.

Robert Szucs is a Hungarian born cartographer and artist who creates beautiful images of watersheds in many continents of the world. On his website, Grasshopper Geography, you can find your own continent and see how the watershed you live in connects you to other territories through the river system that flows through your area.

There are absolutely no contour lines to define the landmass on these maps. The shape of the continent is beautifully described only by the lines of the

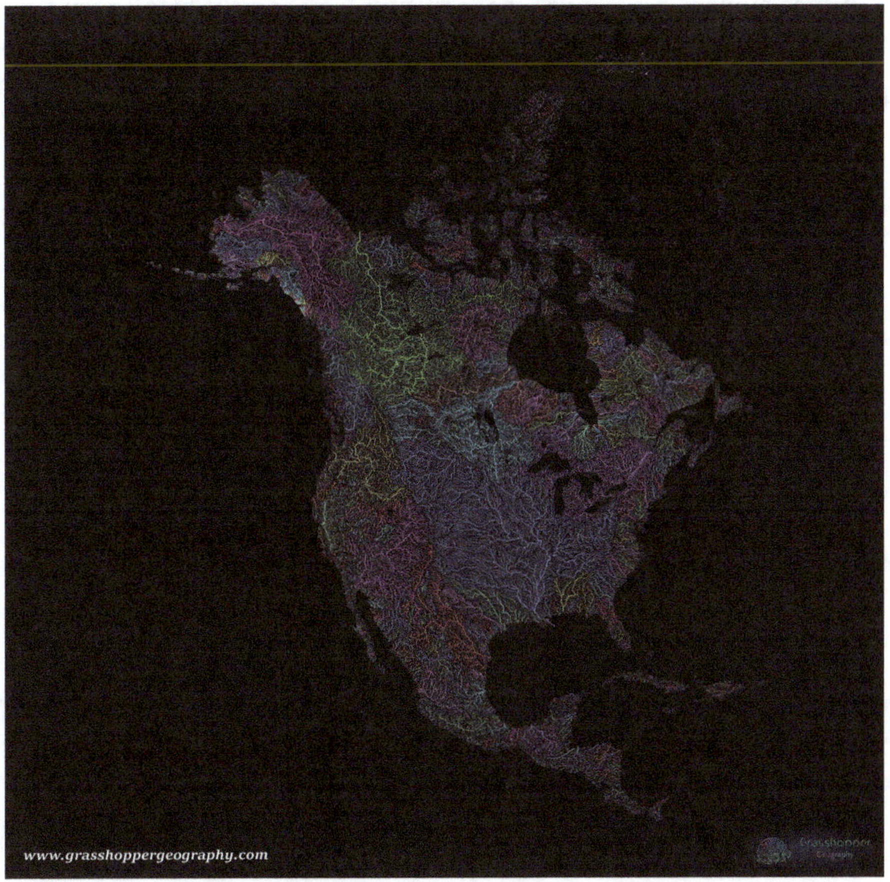

rivers. Just like when you see an anatomy drawing of the human body, the blood vessels describe the shape of the body. When you look at this map of the North American continent, each watershed is defined by a colour. You can see that some of them in coastal areas are quite small, while some of the inland ones are enormous. The Mississippi watershed, in dark blue, covers a third of the landmass of the United States, running from southern, central Canada and empties out in the Gulf of Mexico. Fertilizer runoff from farms in the Central United States travels all the way to the Gulf of Mexico causing dead zones of de-oxygenated water and killing off fish. Through water we are connected. How we manage the land affects the water and those consequences travel downstream, affecting life further down the river.

Just as humans have found that wetlands were inconvenient, they have also decided that rivers take up too much space. Although we love to live near a river, we don't want to lose any space around it. We want to grow crops, landscape our property right up to the riverbanks and build houses as close as possible to the river. But then we find it inconvenient when the river floods its banks, flooding into our basements and putting our crops underwater. Sometimes, especially in cities, humans have tried to impose their will on a river and straighten its course, just because it is hard to build around a winding river.

A river wants to meander

The Nature of a river is to meander. It is a matter of natural water dynamics. When flowing water encounters an obstacle, it deviates its flow to go around it. When water goes around a curve, the velocity of the flow is greater on the outside of the curve than it is on the inside of the curve. This difference causes the river to deposit sediment on the inside of the curve, while the current cuts more rapidly on the outside. Over time the sediment deposits can create little islands which further deviate the path of the river.

Water flows shift with the seasons as well. Sometimes the stream almost dries up and at other times there is an unusual amount of water that flows in the riverbed. This is all part of Nature's system to regulate water flow over time. A natural river can deviate its course through the riverbed quite dramatically over years. Just as wetlands serve to slow the flow of water over the landscape, provide flood mitigation and filter sediment, rivers provide the same functions in their own way. A straight river will flow faster than a meandering river and is much more likely to overflow its banks. A meandering river with its little pools, curves and offshoots will contain the times of heavy flow and allow the water to sit and infiltrate into the land when it wanders through.

The flow is faster on the outer curve and slower inside. A river meanders and changes its course over time.

We realize more and more that water quality problems are being caused by the way we interfere with rivers and take up their space by settling too close to their banks. So new regulations are now being created in many places to protect rivers. For example, Riparian buffer zones are either wild or planted areas along the banks of rivers. Vegetation growing on river banks helps to filter out contaminants and sediment that would otherwise get into the water.

Coastal ecosystems: where land meets water

Coastal ecosystems occur in areas where land meets water, where freshwater meets saltwater and where shallows transition into the deep. They encompass all land areas around the coast: beaches, ponds, marshes, estuaries where sea waters mix with waters from the river mouth, and shallow parts of the seacoast, such as mangroves, coral reefs, kelp forests, and beds of seagrass.

Coastal ecosystems, just as rivers and terrestrial wetlands, have been vastly undervalued and degraded by human presence. Yet, they are an inestimable resource of biodiversity and ecosystem services. A third of the world's population lives along the coasts, even though coastal areas represent only 4% of the land mass. We have destroyed these valuable ecosystems, clearing them away to make room for our homes, farms and infrastructure, or polluting them with agricultural and industrial chemicals. Waterfront property is the most valued all around the globe.

The most obvious resource we get from coastal ecosystems is fish. Fish has been an important protein source for humans since time immemorial. Yet fish stocks all over the world are being depleted, fisheries are now seeking fish species lower down the food chain, and at deeper levels of the ocean. The coral reefs and shallow marine ecosystems provide the habitat for many fish species to lay their eggs in and to protect their young. By failing to protect and conserve these areas we are destroying what is of obvious value to us in terms of both food and climate regulation.

Another important reason to preserve and restore coastal ecosystems is because such ecosystems protect the coast from storm surges. In this time of sea level rises and increasingly severe storms, the buffer zone provided by shallow ocean vegetation such as mangroves, kelp forests, sea grasses and coral reefs, breaks the force of big waves, protecting both lives and infrastructures. In the same way that inland wetlands and properly managed riverbeds mitigate flooding, coastal wetlands do the same at the coasts. Inland salt marshes, estuaries, and places where the excess water gets captured and infiltrates slowly into the land and air, form important lines of defense against flood and erosion. If we look at the costs of recovering from flood disasters in coastal communities, and compare that cost to the expense of conserving and restoring these buffer zones to mitigate disasters, investing in mitigation begins to look good.

All of these places which buffer the land against storms also capture sediments and avoid them from flowing further. This is why they are also rich places, because the trapped organic matter sinks and decomposes, releasing the nutrients they carry. This organic matter provides food for aquatic plants, insects, fish, birds and invertebrates, making these spots treasure troves of biodiversity. We have already talked about the value of biodiversity, which ensures the maximal functioning of all types of ecosystems.

The trapping of organic matter, which in turn leads to a flush of fertility and growth, has another value which is just starting to be recognized. Organic matter is carbon. The new climate change buzzword is sequestering carbon. Scientists have only now recognized that coastal ecosystems sequester more carbon than terrestrial forests: a lot more, in fact. Sequestered carbon is organic matter that is stored in the earth and does not off-gas: it stays in the ground and is not returned in the atmosphere. When decaying organic matter is piled up and trapped under water, it continues to accumulate there for very long periods of time. Some of the carbon that is counted when we evaluate it for carbon credits is the biomass of the vegetation that grows, but the carbon stored in the soils below the vegetation is also of value. In the case of underwater ecosystems, the amount stored in the soil represents 80-90% of the total value. This is because layers of sedimented organic matter continuously build up on top of one another since there is no air to make them compost.

Blue Carbon

The new term for carbon sequestered by coastal ecosystems is "Blue Carbon". All coastal ecosystems are a valuable source of blue carbon but mangroves are the richest, with the capacity to sequester as much as 10 times more carbon than terrestrial forests. But mangroves have been disappearing steadily due to human intervention. The number of intact coastal ecosystems remaining on the face of the earth is negligible. As much as these ecosystems have the potential to sequester carbon, each time one is destroyed, hundreds of years of stored carbon are released as emissions.

Mangroves are one of the richest ecosystems on Earth. They protect coastlines from storm surges, provide habitat for numerous valuable aquatic species and sequester more carbon than terrestrial forests. And yet humans have destroyed the vast majority of mangroves all over the world.

Coastal ecosystems exist on every continent, with the exception of Antarctica. The potential for restoring them as a form of climate mitigation is great. The notion of Blue Carbon holds much promise as an incentive for their protection. In Cispata, Colombia, a consortium of organizations, from non-profits to research agencies and Foundations, has come together to turn the conservation of the mangroves there into an investable proposition through the carbon market. Verra, the organization which sets standards for carbon measurement and reviews projects investable for credible carbon markets that may be eligible for investment in the carbon market, has just registered the Cispata mangrove conservation as its first Blue Carbon programme. Hopefully it will be the first of many initiatives that create incentives for countries, investors and communities to protect and restore these precious natural assets.

Tree cover and forests

We have already discussed the vital role that trees and forests have in the small water cycle. Deep rooted trees penetrate into the earth, keeping the soil permeable and allowing water to infiltrate to deeper layers. The shade of trees prevents the water from evaporating quickly into the atmosphere. Trees also have a critical role in stimulating rain cycles. It is a well-known fact that when regions become deforested, they get less rainfall. The earth's surface was once covered in large forested areas. Today only 34% of Western Europe is forested compared to 80% 2000 years ago. Half of the forests are gone from the eastern North American continent and China retains only 20% of its once vast forest cover. Humans have cleared forests around the globe to make way for agriculture to exploit the timber.

Greater awareness in the developed world has slowed the rate of deforestation. But in the tropics, the rate of destruction of the rainforest is alarming. Tragically, when the land is cleared for agriculture, particularly in the tropics, it does not remain fertile for very long. When soil – originally covered and protected — is exposed to the elements, when humans use bad agricultural management practices, and when rainfall decreases, deforestation is the pathway to desertification.

We cannot afford to lose more forests if we want to save ourselves from an escalating climate crisis. It is imperative to find ways to produce food for inhabitants where natural ecosystems have already been sacrificed.

Chapter Four
Green Infrastructure and Earthworks

We have been talking about the value of natural ecosystems and the ecosystem services they provide. Having devastated – through our ignorance – most of the natural ecosystems that used to cover the surface of the earth, we now face the consequences of the imbalance created by this loss. The parallel crises of the carbon cycle, the water cycle and the alarming loss of biodiversity are all caused by this disruption of ecosystem functions at scale. It is a gross oversimplification of a complex story if we think that we just need to stop burning fossil fuels in order for the atmospheric carbon to drop to pre-industrial levels. We need now to restore the carbon cycle, the water cycle and biodiversity all at the same time. To do so, we not only have to conserve the natural ecosystems which remain, but we need to create functional ecosystems over more of the earth's surface.

Happily, there are ways to design systems that respect and imitate Nature. Indigenous peoples all over the world knew how to, using the resources of the land to enable Nature to continue to thrive. By observing and understanding the ecological functions of natural systems, we can design cities, farms and even industrial spaces so that they meet our needs and also re-establish ecosystem services.

This is the basic premise of the regenerative movement. How do we make our human systems replicate natural systems so that we can restore the healthy function of the biosphere? The knowledge of how to do this is already here. Now it is time to put it into practice at scale and fine-tune the methods.

Let's turn back to our focus on restoring the water cycles and look at the knowledge already available on how to build human spaces which can bring back the small water cycle for our benefit. Although this knowledge is available, serious re-thinking is needed because the current standard of water management systems is short sighted and unsustainable.

Our primary task is to maintain the small water cycle in order to rehydrate the earth. There is a finite amount of water on the planet and the issue is in how it is distributed. Modern management practices have focused on draining, which feeds the large water cycle, moving out to sea. Sea levels are rising, ostensibly because of glacier melt, but perhaps also because we are draining water from the land and letting it run rapidly towards rivers and streams. As temperatures warm, evaporation increases, rainfalls are more erratic and extreme. Heavy rains cause flooding and erosion as they run through drainage infrastructure, carrying topsoil, organic matter and contaminants.

Every day, massive quantities of water are pulled out of lakes, rivers and wells to be used for drinking water, household use, irrigation and industry. Our current thinking on sustainability is focused on water conservation. We have toilets that use less water, washing machines that use less water, irrigation systems that use less water, and we have attempted to recycle used industrial process water. More efficient water use is important, but it does not regenerate the water cycle: it merely slows the decline.

The above approach is similar to what we are doing with the carbon cycle. The full solution is not only to reduce emissions, but also to re-establish the carbon cycle. At present we are pulling carbon out of the earth and sending it into the atmosphere faster than the carbon can return to the earth. If we want to balance the cycle it means not only reducing emissions but also increasing drawdown with natural vegetation. The amount taken out of the bank has to be equal to or less than the amount deposited.

Similarly with the water cycle, we cannot continue to use massive amounts from the groundwater reserves if they are not being replenished at a rate equal to or greater than the withdrawals. **We therefore need urgently to go beyond water conservation and think about how we can refill the aquifers and store water in the earth. It can be done, but everything needs to be re-designed according to regenerative principles.**

Slow it, spread it, sink it

The regenerative principle at the core of restoring the small water cycle is: "Slow it, spread it, sink it." The way that water gets from the sea back to the land is through rain. We have talked about rain cycles and the role of vegetation and trees to bring rain inland. Once rain falls, the way to recharge the small water cycle is to make sure it sinks into the ground instead of running off. The faster it runs off, the faster it leaves the land on which it fell. The principle of slow it, spread it, sink it, means the longer each drop of rain stays in the ground where it fell instead of moving on and going somewhere else, the better.

Think about a bathtub. Imagine it fills up to the brim. Then imagine how quickly it drains when you pull the plug and let the water run down the drain. Now imagine that instead of pulling the plug, you let the water run over the brim. The surface area to where the water spills is much wider than the drainpipe; the water will spill more slowly over a larger area. Then imagine if instead of landing on a hard bathroom floor, it was landing on a porous surface like fertile soil: most of the water would get absorbed instead of running away. This is exactly what we want to achieve.

As water flows down a slope, we want to put barriers in its pathway which will stop it from flowing immediately to the lowest point. These barriers allow the water to accumulate over a larger area. Think about the bathtub. As the water collects in a spot, it spreads over a wider area. It continues until there is too much to be contained there and then it starts to brim over. But if there is no one low point for the water to flow out of, it seeps out like water going out over the brim of the tub, over a wide area. In addition, the water sitting still in one spot or moving very slowly, is gradually seeping into the ground and evaporating into the air. So we are hydrating both the air and the land at the same time.

This is the principle of slow it, spread it, sink it. We slow down the flow by blocking the waters pathway. We give the water a wider surface area to spread into. And we allow the sitting water to infiltrate into the ground very gradually. It seems very simple but the effects of these techniques can make a powerful and significant change in the landscape.

Keyline design

In order to put the principle of slow it, spread it, sink it to work, we first have to understand how water flows over the landscape. It always flows to the lowest point. The greater the slope, the faster it flows. Where the ground is level the water sits still. On a gradual slope, it flows more slowly.

Percival Alfred Yeomans was an Australian inventor who developed what is known as the keyline design system. He wrote a book called Water for Every Farm, which was published in 1954 and influenced many thinkers around the world including many in the permaculture and soil conservation movements.

Water flows down from the highest ridge. The keypoints are the highest indentations where water would collect. Keylines are the contour lines where the land is at the same elevation. If we trap water at the keypoints and dig ditches along contour lines, water will travel more slowly down the slope and sink into the landscape.

The essence of keyline design is to understand the shape of a piece of land in terms of how water will land and flow over the property, and to manage the flow in order to maximize its absorption into the land. Water will fall to either side of a ridge or high point. There will be undulations in the shape of the land, water will accumulate in the valleys and flow off the ridges. If you want to slow the flow of the water, you need to determine a keypoint. A keypoint

is the first spot near the top of the main ridge where water will naturally go to and accumulate. A keyline is the contour line which is exactly at the same elevation as the keypoint.

There are two possible ways you could proceed, depending on the overall design of your land. You could dig a deeper hole at the keypoint and dam it, so that you have a pond uphill on your land. Water from an uphill dam would flow by gravity to lower points where it might be needed.

But if it's not convenient to put a pond at that spot, you could dig a swale (a shallow ditch) or a trench along the keyline. The keyline is the contour line which is at the same level as your keypoint. If you dig a swale along that contour line with a slight berm (a raised bank) or mound on the downhill side of it, water will accumulate in that swale when it rains hard. If the swale is shallow, the water may not stay for very long. It will be absorbed into the land fairly quickly and the swale and berm can be shallow enough

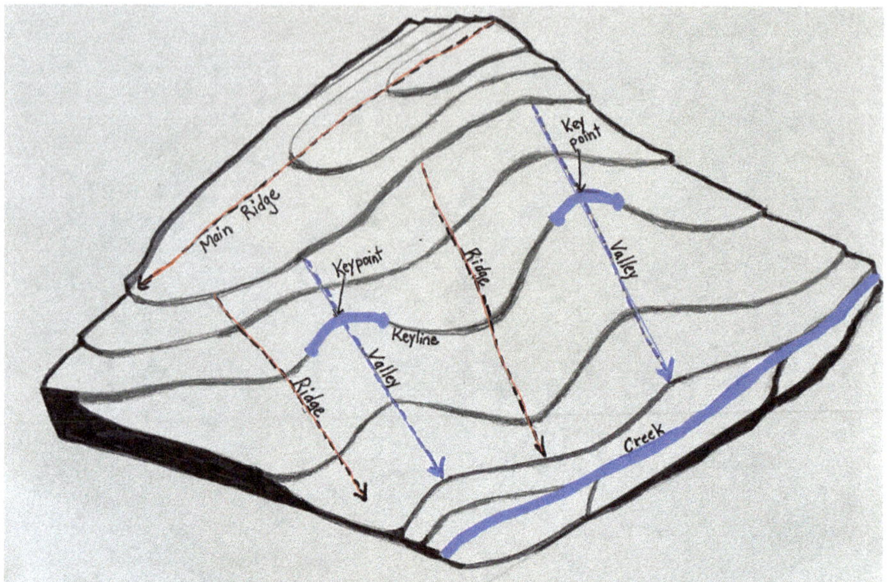

that you could mow right over it. It doesn't have to significantly alter the use of that part of the land.

You could also make the swale with a gradual slope to deviate the water that it collects to another keypoint. There you have the same options. You could make a collector pond at the lower spot, or you could just make a shallow hollow spot and another swale to deviate the water in a zigzag pattern down the slope of your land.

Any one of those key points could be the site of a small or a large pond, depending on your needs. If you have large rainfalls at some points in the year, followed by dry periods at other times, there is an advantage to collecting a reserve of water when it falls in a pond or even a series of ponds. The swales which deviate the water across the landscape can be deeper or shallower. If they are deeper with a very low grade slope, or no slope, they may contain water for a longer time. Sitting water permeates slowly into the land, hydrating it and feeding the groundwater reserves. Even if you don't make any ponds to collect rainwater for a longer storage time and your swales are just shallow undulations, you will still slow the flow of rainwater and make more of it sink on your property. The more carbon you have in your soil, the more of this rainwater will be stored just in the ground, even without ponds.

The core principle behind keyline design is really to be aware of the contour lines which define the structure of a piece of land and determine how water lands on and flows through the property. By making water soak in on the keylines, we can really decrease the speed at which water traverses. P. A. Yeomans also designed a special plow which rips a deep furrow along a keyline. Without needing to dig with an excavator, by keyline plowing, a farmer can make more water sink deeper into the field, because the water flow downhill will be slowed simply by having deeper penetration along the contour lines.

The addition of dams and ponds may not always be needed. But if you live in a place where water is scarce and often comes heavily at only a particular season, collecting reserves from the rain, rather than drawing it from groundwater can make all the difference to the sustainability of water use in that place.

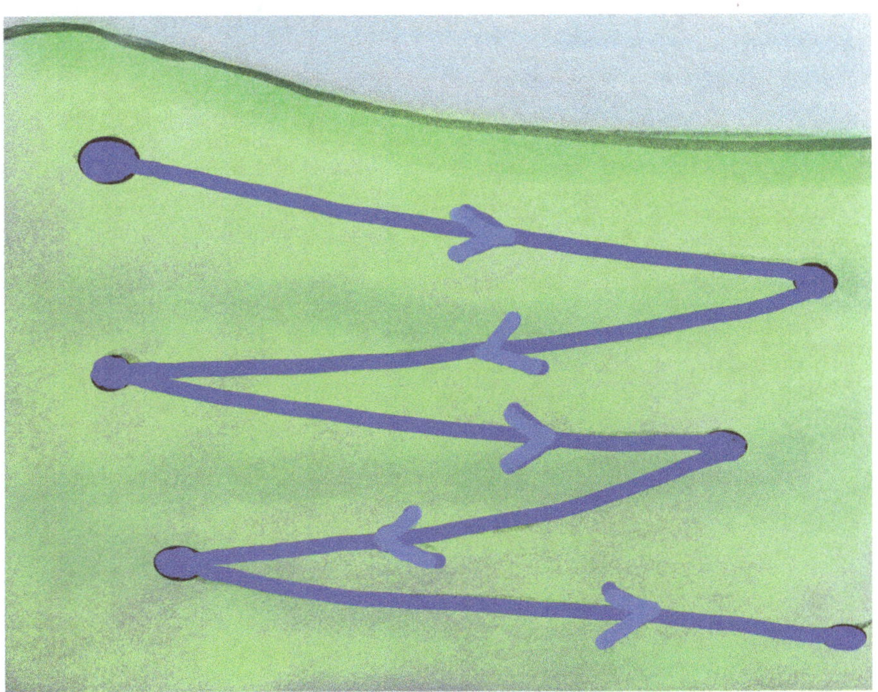

Chapter Five
Bringing degraded land back to life

The story of Coen Farm

Takota Coen is a young farmer and consultant in Alberta, Canada. He farms a 250 acre property with his parents at Coen Farm, in central Alberta. Takota's family has farmed this land for four generations, but Takota wanted out of farming when he finished school. After several years of exploration and discovery however, he decided to return to the farm in 2012, because

he had come to believe that farming and the food system was not only the source of many serious environmental problems, but also the potential remedy. He wanted to restore the land by putting into practice what he had learned about permaculture and regenerative agriculture.

Takota has a photograph of his great grandparents on their homestead taken in 1914. At the time, a hand dug well supplied all of their water needs. The well was only fourteen feet deep. The farm borders Red Deer Lake, which in 1914 was a spring-fed lake. During the one hundred years before Takota returned to the farm to restore the land to its true potential, water levels had dropped drastically and adequate water supply had become a serious issue for Alberta farmers. Although Takota's parents were environmentally minded and had converted the farm to organic in 1988, water issues remained very problematic.

Red Deer Lake no longer had springs. Wetlands which had once been on the property, home to beavers, muskrats and abundant other forms of life had been filled in or had dried up. The well which supplied the farm was 180 feet deep and the family had to trickle feed water all night to fill a reservoir in the barn to provide the required amount of drinking water for the cattle.

There were no more trees left on the farm. In 2002, a serious drought in Alberta meant that barely 2 inches of water fell in the space of a year. Wildfires burned out of control and fire spread underground because there was so little moisture in the land.

Takota's plan was to improve the soil by switching to holistic managed cattle grazing, by planting trees and perennial shrubs in agroforestry plantations, and by restoring the water cycles using the keyline design and the principles of slow it, spread it, sink it.

He researched the ecosystem which had formerly thrived in his geographic area prior to colonization. The original biome was a combination of grasslands with patches of tree cover. He looked for where there had formerly been wetlands. And he studied the contours of the property to understand the water flow patterns.

Double D dam collects runoff and snowmelt from the highest point on the farm. By opening a valve, Takota can choose to run some of the reserve into the lower swales to hydrate the land, provide drinking water for cattle and irrigate the agroforestry plantations.

The design finally conceived and implemented involved digging three major swales about 3 feet deep and over 600 metres long at different levels of elevation. The highest swale feeds a large dam near the top of the property which holds 30,000 gallons of water. Fifty percent of the annual precipitation comes in the form of spring run off when the snow melts. That is why it is important to take advantage of this large quantity of water which comes only once in the year. This dam has a valve which can be opened or closed to allow the water to flood into lower swales and into a series of smaller ponds, dams and swales which wind their way through the land. All together there are three kilometers worth of swales directing water to strategic locations.

The dam holds a reserve of water to provide the farm with resilience in case of multiple drought years. Takota figures he has enough water stored to provide the farm's need for 5 years and he has plans to add more structures to increase that reserve to 10 years.

When he opens the spillway from Double D Dam to flood the lower earthworks, the ponds and swales stay full for weeks as the water infiltrates slowly into the ground. The water feeds into berms on which the fruit and nut trees have been planted and fills the system which provides drinking water for the cattle. There are taps installed every 200 metres throughout the farm to provide clean drinking water in all of the different paddocks where the cows graze.

The farm is home as well to pastured pigs. The constructed wetlands are the perfect habitat for the annual growth of perennial cattails. By July once the wetlands have almost dried up, the pigs are moved in to feast on the year's crop of cattails, which happen to be highly nutritious.

It took 100 years to degrade the land, but this type of intensive restoration is bringing the water levels back much more rapidly. In 2015, Alberta had another major drought and following on the heels of that came the worst flooding that had been experienced in decades. Although the neighbouring farms suffered major hardship from both of these events, Coen Farm did not. They had the water reserves to meet all of their needs through the drought, and their soil and the water infrastructure absorbed the excess water from the floods without any problem.

The wetlands are staying wet for longer as the years go by; the springs are returning in Red Deer Lake; and the well is providing a much faster debit of water than ever before. This change took less than a decade. As the Coens increase the amount of carbon stored in their soil through regenerative agriculture practices, the land is able to store greater and greater amounts of water. One real surprise was when their neighbours came on a tour and heard about what the Coen family has been doing, they realized that their efforts were actually recharging the wells of surrounding farms and not only their own!

The story of Mulloon Creek

Mulloon Creek runs for 50 kilometers along a valley in New South Wales, Australia. It is a part of the watershed catchment of the Upper Shoalhaven River which is an area of about 400 square kilometers. Mulloon Creek became the site of a revolutionary and innovative restoration project which now involves the collaboration of a growing number of people and organizations.

But none of it would have happened had it not been for the coming together of two exceptional men in 2005. Peter Andrews and Tony Coote had a common passion and complementary skills, without which this project would never have been possible.

Peter Andrews is a farmer, a breeder of champion horses and now a consultant on repairing degraded landscapes. He grew up on the land, learning to "read the landscape" from indigenous mentors. He saw the degradation of Nature exacerbating over time and worsening with the impacts of climate change. His early experience led him to an intuitive understanding of how the ecosystem of the Australian landscape used to function and what it would take to restore it.

Colonization, with the European methods of grazing and agriculture, destroyed the hydrology of an ecosystem which had always, despite long periods without rain, been able to store enough water in the landscape to sustain an indigenous vegetation adapted to the natural wet and dry cycles. Before European settlement, the Australian landscape contained numerous wetland areas, little ponds, streams and creeks which could store enough water in the ground, to keep the vegetation alive, even if the streams dried up for some time.

In the European model of farming, wetland areas were overgrazed, causing erosion of the riverbanks. Consequently, topsoil was washed away so that creeks and rivers began to flow more quickly during wet periods, because the vegetation which had slowed the flow of water was lost. The lack of soil cover in floodplains led to the land drying up during drought. Adoption of non-native crop species requiring high doses of agricultural chemicals lead to further soil degradation. This in turn led to increased droughts due to loss of soil fertility, and loss of vegetation. When the rains did come, the eroded river banks and cut deep into the landscape like etched-in crevasses. The water, when it comes, flows away quickly, no longer spilling into the floodplains and feeding the wetlands. Many of these creeks and rivers dried up completely leaving the landscape extremely vulnerable to wildfires, as was painfully evident in 2019.

Peter honed his methods for repairing dehydrated landscapes in the 70's on a degraded 2000 acre grazing property he bought, called Tarwyn Park. Tarwyn Park was a highly eroded and salinized piece of land when Peter acquired it. It became his living laboratory for testing his methods, methods which were unorthodox and controversial at the time. Two aspects of his work led to vehement protests from many of his neighbours. One key part of Peter's strategy is to block the flow of the creek or river in order to slow down the flow, along the lines of the slow it, spread it, sink it philosophy. The neighbours were sure that if Peter blocked the flow on his property, it would mean less water coming downstream to them. The other thing that Peter did which upset the neighbours and the local authorities was to use invasive weed species such as willows, reed plants and blackberry canes in order to kickstart the growth of vegetation on the barren riverbeds. The conventional wisdom of the time was that invasive species needed to be pulled up and it was actually illegal to plant them. But Peter used them because they would take hold more easily on inhospitable ground than other plants could. He would then slash them down, following the *chop and drop* method, creating a mulch of organic matter on the ground. It could then harbour moisture and protect the other species, which would gradually begin to grow once the increased groundwater and increased organic matter made the ground more fertile.

The dramatic revitalization of the land at Tarwyn Park proved that Peter's unorthodox methods worked. He began to build up some credibility and started consulting work for other farmers. Peter called his methods Natural Sequence Farming; it included some basic principles for re-establishing the pre-colonization type of balanced ecology, such as hydrating the land and restoring natural nutrient cycling.

Tony Coote was a successful businessman with a passion for the land. He purchased a farm that borders 3 kilometers of Mulloon Creek in order to pursue his interest in regenerative natural farming. That Farm is still operating today under the name Mulloon Creek Natural Farms. In 2005, Tony Coote invited Peter Andrews to his farm to share his theories and experience on the restoration of watersheds. The encounter was a true meeting of minds and the beginning of a successful partnership. Peter had the intuitive genius and Tony had the people skills and the organized business mind. And they both had the passion. Tony said of Peter, "He can look at a landscape like a musician can look at a piece of music."

In 2006, during a serious drought, Tony hired Peter to design a landscape restoration project on his farm. Tony brought in the Southern Rivers Catchment Management Authority to help manage and monitor the project and he obtained some funding from the National Landcare Program. In this way he received some support from national authorities as well as credible scientific data to justify the project as a pilot study. By doing so, he managed to avoid some of the kickback that Peter had gotten with his cowboy approach at Tarwyn Park.

The project involved fencing off the riparian zone the soil – to protect it from livestock and wild creatures, raising the water level of the creek by installing 16 erosion control structures at intervals along the creek, and planting thousands of trees, shrubs, reeds and rushes in the floodplain alongside the creek. The erosion control structures were each unique in their construction to suit the particular topography, but the general principle was the same. Peter calls these structures "leaky weirs."

Leaky weirs are similar to dams but are not intended to stop the water flow, only to slow it down, allowing the water to infiltrate into the surrounding area and trickle downstream more slowly. In order to raise the water level in the creek, where the banks were typically cliff-like with a sharp drop off, rocks, gravel or logs were placed into the creek. The banks were further reinforced by planting vegetation. Stream gauges were installed upstream and downstream of each structure to collect data on how the weirs affected the water flow upstream and downstream of each structure.

Construction of a "leaky weir" on Mulloon Creek in spring of 2018

The same location on Mulloon Creek in autumn of 2020

During an extreme drought, results from the first year showed that even when no water was flowing from upstream of the farm, there was always water flowing out downstream from the project. The same amount of water flows, but it is spread out over a greater area and slowed down so that it sinks into the floodplain. This allows the vegetation to thrive and a diversity of flora and fauna to establish itself along the floodplain.

Ten years later, fish, birds and frogs – species which had long disappeared – are happily established in the lush habitat. When rains come the water recharges groundwater, which act as a reserve for the drought periods. When the rest of Mulloon Creek dries up, this section of the creek stays green. The complexity of the habitat generates more organic matter and nutrient cycling. Farm productivity has increased by 60%. The water quality is improved and the water flow is more constant.

In 2013, Tony Coote and his wife established the Mulloon Institute in order to raise awareness, to scale up the project and to carry out well documented research on regenerative farming. Twenty more landholders were brought into the project, increasing the scale of the work to 23,000 hectares along a 50 kilometer stretch of Mulloon Creek.

Sadly, Tony Coote passed away from cancer in 2018. He was mourned by many people who recognized the huge importance of the work he had started. As a wink to the detractors of Natural Sequence Farming, Tony was buried in a casket woven out of natural willow branches.

Despite the loss of Tony, the work he began continues to expand and gain momentum. Peter Andrews is still doing his consulting work. The Mulloon Institute continues to monitor and collect data on the obvious success of the expanded project, now called the Mulloon Rehydration Initiative. The objective is to become a model which can be replicated all over Australia. In 2019, Scott Morrison, the Australian Prime Minister visited Mulloon Creek, along with David Littleproud, the federal minister for Agriculture and Water Resources. The Mulloon Institute was subsequently awarded a grant of 3.8 Million $ over five years to continue the scientific studies and to train land managers in natural sequence farming, regenerative agriculture and holistic management.

The story of Al Baydha

The Al Baydha project is a remarkable story of restoration that is taking place in rural Saudi Arabia, in the foothills of the Hejaz Mountains, south of Mecca. The project was conceived in 2009, when Saudi Royal Princess Haifa Al Faisal visited the region and became concerned about the extreme poverty of the residents. The area is inhabited primarily by Bedouin tribes, formerly nomadic peoples who lived by herding camels and goats. In the 1950's their traditional indigenous ways were banned by the Saudi government and the Bedu were forced to remain in place. The consequence of stopping the movement of these tribes with their herds of animals was that the land became quickly overgrazed and desertified. When the people couldn't survive from their herds anymore, they were forced to cut down trees and shrubs to sell or provide for their own energy needs. Over time the land became more and more barren.

Although the area is naturally a desert, there was, just like in Australia, an adapted desert ecosystem which had supported this indigenous lifestyle

The Al Baydha site in 2011 at the beginning of the work

for centuries. Like in Australia, reverting to unsustainable grazing methods quickly degraded a fragile landscape which had managed to adapt to scarce annual rainfall and erratic rainfall patterns. Residents of the region recall that there were once large trees on the landscape. But by 2009, there were only rocks, hills, intense sun and dust. The people were severely impoverished and the regional economy was crippled. The average annual rainfall is 60mm and summertime temperatures often reach 50°C.

Her Royal Highness set out to find a way to turn this problem around. She reached out for help to figure out a way to re-establish a sustainable economy for the region, and contacted Mona Hamdy, a bioethicist and futurist from Harvard University, and Neal Spackman, a permaculturalist from Stanford University. The official project began in 2010, with the founding of the Al Baydha Development Corporation under the direction of Mona Hamdy. The primary goal was to create a viable and sustainable financial and social economy for the more than 3000 tribal Bedu inhabiting the 7000 square kilometer region by employing them and training them in building and managing the project infrastructure. The environmental goal of reversing desertification was aimed at improving the economic and social dilemna. Neal Spackman was hired and came to live in Al Baydha from 2010-2018 to manage the landscape restoration part of the project.

The geography of the region is typical of the watersheds of the entire west coast of the Arabian Peninsula: a desertified mountainous area which spills onto a narrow floodplain before hiting the Red Sea. Because of the slope and because there is no vegetation to absorb and retain waterfall, over 90% of the water that falls as rain flows rapidly down the slope and is lost into the Red Sea. When the rain comes, it tends to come in a flood, washing away any soil that may be there, along with bits of debris and organic matter. Neal's job was to design and implement a plan to restore fertility to this barren landscape, along with the crew of Bedu workers.

In consultation with fellow permaculture expert, Geoff Lawton, they designed a low-tech passive system to slow, spread and sink the flood waters, trapping them in the landscape and capturing as much of the available water as possible before it left the floodplain and dissipated into the sea. The crew of Bedu workers moved rocks to create check dams, blocking the flow of flood waters at key points on the slopes, and created a series of level terraces along the mountainsides to slow the flow of water and encourage it to sink into the land.

For three years the work focused on building these structures to trap the rain. It only rained four times in those three years, but some of those times it rained so hard that some of the structures were washed away. The team had to go back and re-design a new structure that would be sturdy enough to withstand the pressure of intense floods.

By 2012, the work shifted to planting trees. The trees were intended to stabilize the landscape and sink water in more deeply. But also the long-term vision of the project was to build an economy based on sustainable grazing and agroforestry. Ten species of trees were selected, based on their potential to have an economic or an ecological impact. By 2015 there were over 4000 trees where once there had been none. The young trees were sustained through a drip irrigation system with water that had been sourced partially from a local well and partially from a desalination plant.

By 2016, 20,000 cubic meters of water from external sources had been used to establish the site. But the site had subsequently caused 50,000 cubic

The same swale as in the previous photo in 2014, after tree planting and a rainfall

meters of water to sink into the landscape. There were signs that the project could be a successful model which could be replicated along the entire Saudi Peninsula. But funding was running out and it was important that this built ecosystem not be dependent on irrigation. In order to truly be a sustainable closed loop system, the vegetation needed to be able to survive the region's climate. So the decision was taken to cut off the irrigation and to see if the system could sustain itself with only the available water resources.

Dramatically, a devastating period of drought followed the decision to cut the irrigation. No rain fell in 2017, nor for most of 2018. The landscape became drier and drier. Trees began to die off. The Bedu implored Neal to bring in water but he refused. The project needed to survive the test of its own climate.

Finally, 27 months after cutting the irrigation and 31 months since the last significant rainfall, it rained in November of 2018. And it rained again in 2019. The landscape burst into life with vegetation. Four of the ten tree species had survived. The site turned into a dryland savannah, with grassland plants and shrubs and a diverse habitat attracting birds and insects. The last step is left to accomplish: to introduce a sustainable grazing system, one that could help the savannah to thrive and at the same time provide a source of food and livelihood for the Bedu population of Al Baydha.

Neal Spackman's hope is that the project serves as a pilot which could be replicated along the 30 million acres of desert coast. If 90% of all rainfall runs off these coasts into the Red Sea, then enough water is lost that could have irrigated 130 million desert trees for three years. Such a massive restoration project would cost billions of dollars, but it could also increase Saudi Arabia's GDP by 3-5% annually, sequester massive amounts of carbon and provide fresh water resources for the area. Such benefits are immeasurable.

Neal Spackman's direct involvement with the Al Baydha project came to a formal end with the production of a video documenting the restoration project. The knowledge and inspiration he gained from this intense and remarkable life experience have led him to continue this path by founding Regenerative Resources, a company whose business is to transform degraded landscapes into circular economies. Neal firmly believes that there needs to be an economic incentive to engage impoverished rural populations. A major driver of much of this land degradation is the loss of traditional lifestyles, forcing people to make choices based on survival, such as harvesting and selling the trees. He believes that it is possible for humans to be a vector of positive change. His business seeks to involve landholders, governments and institutions, encouraging them to invest in projects similar to Al Baydha. He encourages them to work with local, indigenous people to build productive, sustainable landscapes, thus providing ecosystem services for the region as well as a thriving economy for its residents.

The projects include the establishment of dryland agroforestry like Al Baydha project, as well as coastal agriculture, aquaculture and the restoration coastal ecosystems, such as mangrove wetlands. Some funding comes in the form of carbon offsets, sold to companies which invest it in carbon-verified projects such as mangrove afforestation. Mangroves have some of the greatest potential of any ecosystem on the planet to sequester carbon, with an acre of mangroves being worth an estimated $100,000 in ecological services. These restored coastal ecosystems can provide value not only from payments for ecosystem services, but the rich habitats they provide are the basis for productive local economies centred on fishing and other goods which can be sustainably harvested.

Chapter Six
Green Infrastructure in Cities

In the 1950's 30% of the world's population lived in cities. Now that figure has risen to 54%, and experts predict that the trend will continue, bringing two-thirds of the global population to urban areas by 2050. What are the environmental consequences of this migration towards urban centres?

From the perspective of water, cities are vast sprawls of largely impervious materials which displace the natural environments on which they were built. Rivers and wetlands are often removed or altered to make space for buildings. The connectivity of habitat for most non-human species is interrupted by roads and buildings, making the living space and movement corridors of insects, birds, amphibians and mammals fragmented and reduced.

Concrete, asphalt and bricks do not absorb rainfall, so most of the water that falls in cities does not percolate into the ground; nor does it hydrate the earth or refill the aquifers. Anyone who has ever collected runoff from the roof of a building to fill a water barrel knows that even a small roof collects a lot of water. Try to imagine how much rainwater concentrates on all the roofs, roads and paved lots in a large city! Generally, all of this water is collected in the city's sewer system and deviated away into rivers. This water is full of debris and pollutants collected from the surfaces it fell on so it often ends up clogging the sewers and being of very poor quality.

In this time of climate change, with more extreme storms and weather events, the stormwater infrastructure of cities is often overwhelmed. We frequently see photos in the news of flooded cities, with buildings and cars abandoned and high water in the streets.

On the other side of the coin, cities require a huge amount of water to allow all their inhabitants and their industries to drink, wash, flush toilets and feed industrial processes. The water crisis in Capetown, South Africa in 2018, saw the city coming close to being the first in the world to run out of water. After three years of drought, the city's water supply, largely provided by rainfall collection stored in several large dams, was getting dangerously low. The administration decided that if the levels got down to 13.5% of the dam capacity, they would have to call it "Day Zero", meaning that water supplies would be cut off and citizens would have to queue up in the streets to collect their ration of daily water. The city made a concerted effort to incentivize citizens to conserve water, calling for shorter showers, re-use of greywater (wastewater), no filling of swimming pools or car-washing, and heavy fines for households who exceeded rations. Social media campaigns shared tips on how to use water sparingly. Day Zero came close but was narrowly averted. Subsequently rains came and refilled the dams and so far the city has had a respite, giving them time to rethink their water management strategies. Capetown is looking at a range of strategies to make their water supply more resilient, including halting degradation in water catchment areas.

Cities take their water supply from various sources, including from lakes, rivers, dams, wells and desalinated ocean water. The question is, are we withdrawing more water than is being replenished? In many cases, the answer is: yes. If we continue to withdraw from water reserves while badly managing it, so that more water flows out to sea than returns as rainfall, the system is not sustainable. In many countries around the world, water consumption is higher than the volume of rainfall annually. With climate change, these unsustainable practices will lead even more rapidly to Day Zero unless we radically change the way we manage water in our rapidly expanding cities and towns.

In Beijing, a mega city that has historically drawn its water from underground aquifers, buildings, roads and pavements are beginning to sink and crack. This is because the space previously taken up by groundwater is shrinking, due to the aquifers drying up. City planners in cities such as Beijing, Sao Paulo and Los Angeles, where the dry and warming climate combined with unsustainable water management threatens a similar scenario as Capetown, are being forced to think about innovative strategies to better manage water resources to supply their needs

Traditionally city infrastructure for water has been grey infrastructure: pipes, sewers, pumps and valves. But some leading cities are starting to innovate with green infrastructure. Green infrastructure is Nature based infrastructure which imitates natural systems using soil, vegetation and managed ponds and water channels instead. Cities, with their high volume of thermal mass and low volume of soil and vegetation, create heat islands. Green infrastructure can address the problems of water scarcity, flooding, water quality and heat islands all at the same time.

Basically, green urban infrastructure applies the principles of slow it, spread it, sink it to an urban context. Instead of rapidly flowing water going into sewers and drains and being taken away by pipes and evacuated, water is captured in vegetated areas and allowed to permeate into the ground. When water is filtered through soil, gravel and vegetation; it is purified of debris and pollutants. Because it is clean, it can sometimes be recycled and used for other purposes, such as irrigating gardens or used for washing. Green infrastructure in a city helps to offset the heating from the thermal mass of buildings and roads and provides a pleasant atmosphere for citizens as well.

Trees and plants, preferably native species, play an important role in absorbing water and adding shade. By introducing soil and vegetation into city landscapes, cities come closer to being a part of the surrounding ecosystems, providing habitat for insects, amphibians and birds. They also can contribute to climate regulation ecosystem services such as carbon sequestration, air purification and of course, water retention and filtration.

Singapore is one of the leading cities using green infrastructure, showing the way for other cities. Originally lacking its own freshwater supply, Singapore has a contract to import water from Malaysia. But that contract terminates in

2060. So back in 1972, Singapore began thinking about water sustainability. A Water Planning Unit was put in place at that time to explore all possible means of achieving water self-sufficiency and the city has continued to innovate and implement new systems ever since.

Two thirds of Singapore's surfaces, including rooftops, parks, medians and roadways capture rainwater and channel it through tunnels into 18 reservoirs. This green infrastructure now provides 35% of the city's water needs. There are plans to eventually have 90% of the city's surfaces providing water catchment. In 2014 Singapore hosted an international event called International Water Week, to showcase water sensitive methods of urban planning, architecture, IT and green engineering which proved that clean, green technology can be great boost also for economic development. Other cities around the globe are paying attention and many are beginning to implement elements of green infrastructure in their city planning. The following are some of the key elements being now integrated into cities.

Rainwater harvesting

Ten millimeters of rainfall over a hundred square metres of rooftop will yield one thousand litres of water. As mentioned above, Singapore is an amazing example of a large megacity which has begun to work towards water self-sufficiency. Rainwater harvesting is a major pillar of its strategy. It requires a smooth surface – often a rooftop, gutters, pipes, a filter to catch debris and a reservoir. It can be done on many different scales.

India is just one place which is suffering from increasing water issues due to climate change. The groundwater is depleted. The pattern of heavy monsoon rains washing away quickly, followed by scorching hot dry months, is endemic in India. In the town of Pune Maharashtra, the citizens who were tired of spending a large portion of their income to poor quality water tanks came together to find a better solution. With the help of a group of social workers and hydrogeologists, they formed a non-profit NGO, Mission

Groundwater, to recharge the groundwater through rainwater harvesting. The community is on the foothills of mountain slopes and traditionally depended on wells, which were beginning to run dry. The plan was to catch as much rainwater as possible and use it to raise the groundwater levels. Building complexes installed collecting systems from their rooftops and terraces. They even dug trenches along the backs of buildings facing the hills so they could divert water flowing down the hill into the channels to recharge the wells. They also harvested rainwater from the edges of roads along the hills. Built channels direct the water into a filter pit which then goes into the wells.

After the first year of the community efforts, water levels came up in the wells, and some which were dry began to hold water again. The group continues to engage citizen scientists to document the effects on groundwater levels and to convince more landholders to allow them to dig contour trenches on the bottom of the hills to collect even more rainwater into the recharge pits.

Permeable pavement

Different materials, such as permeable pavement, pervious concrete, and porous asphalt, unlike conventional hard city surfaces, allow rainwater to percolate into the earth Pervious concrete is made without the fine sand or clay normally used in cement, and this creates air spaces between the larger aggregates, allowing water to get through. This pervious concrete does not have the strength of conventional concrete, but it can be used in many areas which don't need to bear heavy weight.

Other types of alternatives to impervious surfaces include gravel driveways, or patios, sidewalks and terraces made of bricks with stones in between. Some of these new materials are made from recycled glass or different types of resins. All of them serve the purpose of slowing the runoff of rainfall and allowing it to sink into the ground. Depending on the materials, these porous surfaces can allow two to eighteen gallons of water per minute to soak into the ground.

With the increasing concern about flooding in cities due to climate change, more modern cities are beginning to install or research alternatives to traditional asphalt and concrete.

Green roofs

Green roofs are vegetation planted on a flat or slightly sloped rooftop. There are multiple possibilities, from a thin layer of soil containing shallow rooted ground cover plants, to deeper layered thicker systems growing shrubs, flowers or vegetables. The latter involves a building that is solid enough to hold the additional structural load. To ensure good drainage, different layers of soil are laid on top of one another and mixed with gravel, and different densities of soil medium and a membrane ensure that the water doesn't soak into the building.

The city of Hamburg, Germany has developed a Green Roof Strategy in order to create climate resilience. The goal is to plant 100 hectares of green roofs over the next decade. They are a way to counter the heat island effect, and to mitigate the impact of heavy rainfalls: 40% more precipitation is expected in northern Germany in winter months.

A green roof lasts 30-50 years, much longer than a conventional flat roof. They increase the insulation value of the buildings, making them warmer in winter and cooler in summer. They absorb the worst impacts of storm surges as they retain large volumes of water and filter that water through the plants, thus purifying it before it is released into the stormwater system. This minimizes the cost of upgrading the evacuation system. Green roofs increase biodiversity and provide habitat for birds and insects. Some of them become community parks and gardens, providing enjoyable gathering spaces where food can be grown. There are even sports fields among the green roofs.

Imagine city dwellers spending their recreation hours gardening, playing sports and socializing in rooftop parks.

Bioretention gardens and swales

The principle of bioretention gardens and swales in an urban environment is the same as with keyline swales and dams in the earlier rural examples. Water is collected and retained temporarily, allowing it to be slowly absorbed into the earth. But rather than being oriented only to a slope, the water in the urban environment is collected from impervious surfaces such as roads, parking lots, rooftops and terraces. Sometimes these swales are in the median of roadways, or on the edge of parking lots or sidewalks. The trench is filled with layers of compost, soil, sand, and gravel, or sometimes

In a bioretention swale, stormwater collects and is gradually infiltrated through layers of earth, sand and gravel. Whatever is not absorbed is diverted through a perforated pipe to gradually seep into the ground.

recycled synthetic materials, all designed to filter the stormwater. Perforated pipes may be embedded at the bottom to direct the flow of excess water over the larger area.

Sometimes these bioretention areas are not swales but just gardens dug near a place where water runs off. These rainwater gardens perform the same function but in a smaller area. Plants must be carefully selected so that they can tolerate having their roots underwater some of the time, as well as going dry at other times. The garden or swale may retain standing water for a long time, or a much shorter time, depending on the regional climate and the amount of rainfall. Eventually all the water seeps into the ground so such rainwater collecting does not form ponds.

This type of infrastructure not only serves its intended function of reducing the load on municipal pipe systems to prevent flooding but, like any natural ecosystem, it provides many services at once. It allows water to infiltrate and hydrate the earth and refill groundwater reserves; it improves air quality in the city, and provides habitat for diverse small species such as birds, pollinators and invertebrates. It also makes the city more hospitable for people.

Chapter Seven
Decolonizing our relationship with Nature and building a regenerative culture

In many of these examples of degraded land, we have seen that there was once a natural ecosystem where indigenous peoples lived in a way that was sustainable for centuries. When settlers arrived from other places and began to change those indigenous land management systems, it triggered a process of rapid degradation. The change in relationship to the land is not only a question of sheer human numbers but also of attitude. Indigenous cultures have always had a fundamental value system that focused on preserving the resources they depend on. The dominant global market culture thinks about Nature's resources as things to exploit. It thinks of short-term profit at the expense of long-term depletion. In the current economic system, corporations are rewarded to provide gain in each quarter of the year and not to think about where that will lead in ten years.

In his book *Dirt, the Erosion of Civilizations*, David R. Montgomery talks about the origins of agriculture in the Middle East and how the gradual process of deforestation and plowing the land destroyed its natural fertility. Once the land was degraded and could no longer sustain the people, colonizers moved on to other places, to settle and gradually deplete the land in those places. From Mesopotamia, to Ancient Greece, to the Roman Empire, to China, then with European colonialism and the push from east to west of the Americas, humans have exploited natural resources and moved on to colonize other lands and peoples which were still resource-rich.

When I speak of colonization, I speak of this mindset that feels entitled to exploit Nature and other people, to grab what is there without respect or consideration. When the stocks are used up, colonizers feel entitled to move on and take somewhere else. There is no sense of belonging to the land or being responsible for it.

Market capitalism has become the global culture and displaced older land-based cultures. A sustainable culture is not one which uses, plunders and goes on to get somewhere else. We are now hitting the limits of what this planet can sustain and although some dream of going on to establish colonies on other planets, perhaps it is time to rethink how we live on the Earth.

Sustainable implies that we can continue on with the way we do things for an indefinite period of time. But if our current way of living is rapidly burning up the reserves of carbon, consuming the groundwater reserves faster than they are being replaced and driving most other species rapidly into extinction, we clearly can't continue. We need to do more than slow

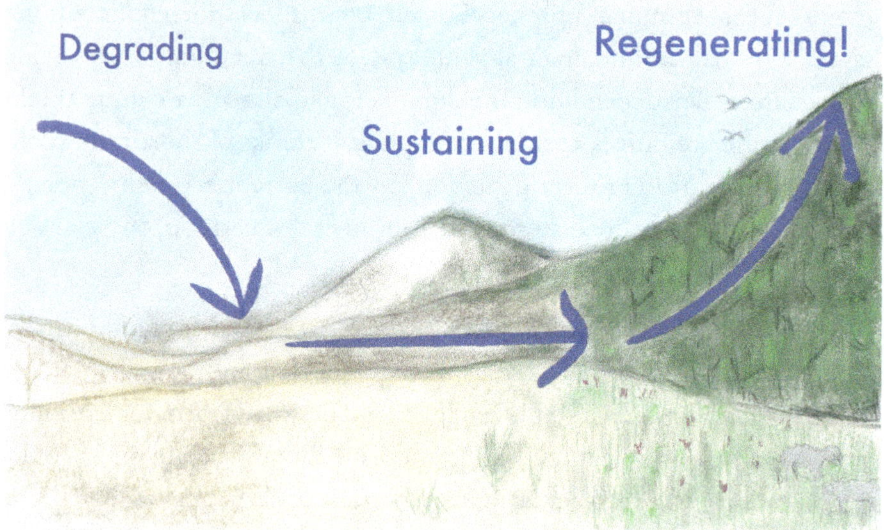

There is no point in being sustainable when the land is degenerated. Sustainable means not getting any worse and we need to do better than that Regeneration means to restore to better than what has been degenerated.

down on this road to annihilation; we must restore and regenerate a liveable planet. To do so, we also need to build a human culture which is regenerative.

What is a regenerative culture? We need a culture capable of bringing all peoples and all countries together in a collaborative effort. There is often a dichotomy between environment and economy. As long as people feel that they need to sacrifice their livelihood to protect the planet, there will be struggle. As long as some groups of people continue to dominate other groups, there will be resistance. A regenerative culture meets people needs and brings humans into relationship with the natural world. It includes all humans.

This may seem utopian, but around the world, initiatives are being taken to restore the land and heal the relationships between communities and Nature. The massive report compiled by the Intergovernmental Science-Policy Platform on Biodiversity and Ecosystem Services (IPBES), released in 2019, contains the contributions of more than 400 authors and tells of the ominous state of health of the global ecosystems upon which our lives depend, showing that they are still depleting at an alarming rate. The report cites the disproportionate contribution of indigenous peoples protecting the biodiversity which still remains on this planet. Although indigenous people make up only 5% of the global population, they steward over 25%of the Earth's terrestrial land surface. These lands represent the majority of the world's remaining intact ecosystems and biodiversity. Without acknowledging indigenous rights to land and engaging their knowledge and collaboration in our efforts to restore our ecosystems, we will fail to achieve what we must to assure our collective survival.

Too often, our idea of conservation is simply to keep humans off the land. This notion is sometimes at odds with indigenous land rights. It is incongruous with our idea of conservation that a human presence can be a beneficial one to the environment. And yet new research shows that low impact stewardship by indigenous and rural populations is often more effective at preserving ecosystems than simply leaving them alone. Kat Anderson studied the traditions of indigenous peoples in California and published Tending the Wild: Native American Knowledge of California's Natural Resources, a book which overturns the notion that native people were hunter gatherers having little impact on pristine, untouched wilderness. In fact, the incredible natural beauty of the landscape was a result of careful tending, selectively harvesting, burning, pruning and enriching the natural resources. Kat Anderson argues that, in this time, we urgently need to learn from these ancient ways and integrate this knowledge into how we manage human systems.

A restoration project which enlists the support of indigenous and local communities and provides them with a means to a reasonable livelihood is the most promising solution. Indeed, with the human population being what it is, the only viable solution is to build a culture based on respect and inclusion, and to reimagine humans as being a part of Nature rather than dominating it.

Around the world there are inspiring examples of people building systems, organizations and projects to change the way we do business. Let's have a very brief look at what some of these initiatives are.

Accounting for Nature

One fundamental problem with the way the global economy works is how we measure value. The money and accounting system of nations is only based on the goods and services that humans produce and consume. The sum total of all these transactions is the GDP of a country and this number is meant to represent the measure of prosperity. As long as we see value only in this way, not only corporations will try to keep their share values up, but also poor and middle-class people will do whatever they can to survive in the short term. **The problem is in a system which gives value to transactions of goods and services, but takes Nature for granted.** The way we see wealth and account for it in the economies of governments around the world needs to change if we are to prioritize restoring and protecting the natural world and the critical ecosystem services which sustain life on earth.

In 2019, the UK's Treasury Department, the department responsible for the UK's financial and economic policy, commissioned Indian British economist and professor Sir Partha Dasgupta to do a review of the economic impacts of biodiversity loss. The final report The Economics of Biodiversity: the Dasgupta Review was published in early 2021 and it has made waves in economic circles. The key message of the report is that our economies are embedded in Nature and the value of Nature must be accounted for in our economic systems.

"Our economies, livelihoods and well-being all depend on our most precious asset: Nature. We have collectively failed to engage with Nature sustainably, to the extent that our demands far exceed its capacity to supply us with the goods and services we all rely on. Our unsustainable engagement with Nature is endangering the prosperity of current and future generations.

"At the heart of the problem lies deep-rooted, widespread institutional failure. The solution starts with understanding and accepting a simple truth: our economies are embedded within Nature, not external to it. We need to change how we think, act and measure success. Transformative change is possible – we and our descendants deserve nothing less."

Professor Dasgupta speaks about Natural Capital, the fundamental assets of functional ecosystems and says that until our economies give financial value to these assets as a part of a nation's wealth, we will continue down a destructive pathway. When we see things in strictly economic terms, if a rainforest is worth much more to a nation staying intact than the profits which can be gained by clearing the land for grazing or selling the timber, then it is easy to make the right choice. The cost of continuing to extract more resources than Nature can regenerate will have severe economic consequences on our economies. It is already happening when we have to deal with climate catastrophes, food insecurity because of floods, droughts or animal-borne diseases.

Between 1992 and 2013 the world economy grew in terms of the capital we account for, but natural capital declined by 40%, according to the review. Unless we change, the time of reckoning when it will be obvious that products and services can't keep growing while Nature declines, will painfully force the system to break. Dasgupta calls on the world's financial systems to account for natural capital, and build in mechanisms such as fees for use of natural resources, a proportion of taxes devoted to the protection of Nature and penalties for the abuse of Nature.

He also calls on citizens to demand this kind of change, recognizing that governments only change because the people demand it. The call for this type of massive change will come from citizens and consumers who are aware of the value of Nature, who boycott those corporations which abuse it, and who invest personally in Nature-based solutions. He cites the example of the concerted effort made in post war Europe to rebuild from ruins, under the Marshall Plan, when citizens and ambitious policies came together to rebuild. Today the monumental effort will have to be focused both on reducing our impact on Nature and on regenerating the biomes we depend on so they can return to productivity.

He cites the enormous cost the world is paying to deal with the COVID pandemic, as opposed to what it might have cost for a decade-long effort to monitor and prevent disease caused by the fragmentation of tropical forests and wildlife trade. Finally, he calls for investments in Nature-based solutions as a critical part of the COVID recovery stimulus. Governments have long invested money in supporting policies and investments which are harmful to the environment; now they should focus all efforts into rebuilding a green recovery.

Many highly influential financial experts came together to write another report which was issued in 2020 by the Paulson Institute, the Nature Conservancy and the Cornell Atkinson Centre for Sustainability. Called

Financing Nature, this report looks into what financial mechanisms are needed to bridge the gap between the funding which is currently available for Nature based solutions and what is needed to restore enough functional ecosystems by 2030. The authors go into details on how the suggestions made by Professor Dasgupta can concretely be put into practice. With the types of financial programs they describe in this report, we have a concrete pathway to motivate people to invest in conservation and restoration of ecosystems.

The evidence shows that a healthy planet is good for business because it will be far cheaper to prevent environmental damage than to deal with it after it has happened. The list of recommended actions goes from measures to avoid damage, to slowing it, to halting it and finally to reversing it. The first thing they recommend is for governments to abolish subsidies which encourage harm such as support for unsustainable agriculture, forestry, fisheries or fossil fuel.

They recommend pathways for both governments and other financial institutions to invest in conservation and restoration. There is a business case for protecting natural infrastructure such as watersheds, wetlands and coastal ecosystems because if we put a value on the services they are already providing for us, we can see that it would be far more costly on municipal and state budgets if we had to replace them. Carbon markets and systems of payments for ecosystem services are being built currently to put a value on natural capital and provide methods to quantify and verify the benefit. Having these systems in place will provide a good source of funding for conservation and restoration and it will also incentivize landholders to keep the ecosystems on their land intact.

There is a growing appetite for Green Financial Products. In fact, there is currently more demand than can be supplied. These include equity investments into green businesses and projects and green bonds issued by governments or financial institutions to pay for conservation projects.

The report goes into another emerging trend in the finance world: risk assessment. This assessment has always been a part of investing, but now it is beginning to include how environmental degradation or catastrophe is a risk to investors. Risk assessment even factors in the negative perception of the public towards companies which degrade the environment. Insurance companies are beginning to look at the damage of environmental catastrophes as risk factors and refusing insurance to some unsustainable businesses.

As a result, companies are increasingly feeling the pressure to improve their reputation by investing in their supply chains to make sure that all the way along the production chain their activities are sustainable, or better, regenerative.

A decade of restoration

The United Nations has declared 2021-2030 as the UN Decade on Ecosystem Restoration. **It is a rallying cry to the world to participate in halting, preventing and then reversing the degradation of all of earth's ecosystems in order to combat climate change, end poverty and prevent a mass extinction.** If all countries and their citizens make a concerted effort to produce results in this next critical decade, we have a real chance at making it a turning point. We still have ten years to turn things around before we reach an irreversible tipping point, according to the scientists who have analyzed the data on climate change and biodiversity loss, even if the present situation is dire.

A large-scale concerted action is required. And we have the knowledge of what we need. There is no time for despair. If we all do what we can to raise awareness and put our individual skills into service, there is hope.

The strategy of the #GenerationRestoration led by UNEP, FAO and partner organizations is to empower a global movement by raising awareness, connecting people and organizations and supporting restoration projects. The Decade of Restoration website invites everyone to make a pledge to support restoration on both large and small scales. Initiatives which have registered to be an official preservation project are displayed on the website.

The Weather Makers Leading Change

The Weather Makers *is a reach for the sky*, massive scale project which gives a lot of hope and inspiration. It is being conducted by a group based in the Netherlands. Using holistic engineering, and working along similar principles as have been described in this book, they aim to stop the devastation of climate change by re-establishing healthy, functional, ecosystems in the driest, most degraded lands on earth. As they explain in a short video, large-scale disruption of functional ecosystems with resulting desertification is the major cause of the climate crisis. As poor land management leaves land parched and barren, more of the sun's heat is captured. It should be noted that today one quarter of the earth which used to be green is now desert.

"An exhausted dry land transfers solar energy into heat. Heat is trapped by greenhouse gases causing higher temperatures, droughts, floods and further desertification. Watershed wide ecosystem regeneration is the basis of every climate change solution: retain water to kickstart ecosystems."

The Weather Makers organization is named after a book by that name written by Tim Flannery and published in 2005. The book shows how mankind is driving climate change and cites evidence that its consequences have been with us for some time. It incriminates the Republican factions of American politics for being in the pocket of the coal industry and for driving climate denial. The book had an important sway on certain influential figures, such as billionaire Richard Branson and Arnold Schwartzenegger, then governor of California, inspiring both of them to take climate action.

The organization was founded in 2017 by three young engineers from the Netherlands: Ties van der Hoeven, Maddie Akkermans and Gijs Bosman. Ties, whose background was in dredging, had worked on the construction of the artificial islands in Dubai, and was contacted by the Egyptian government concerning a project to restore Lake Bardawil. Lake Bardawil is a freshwater lake spanning 400 square kilometers at the edge of the Meditarranean coast on the Sinai Peninsula. It has been an important resource for the local

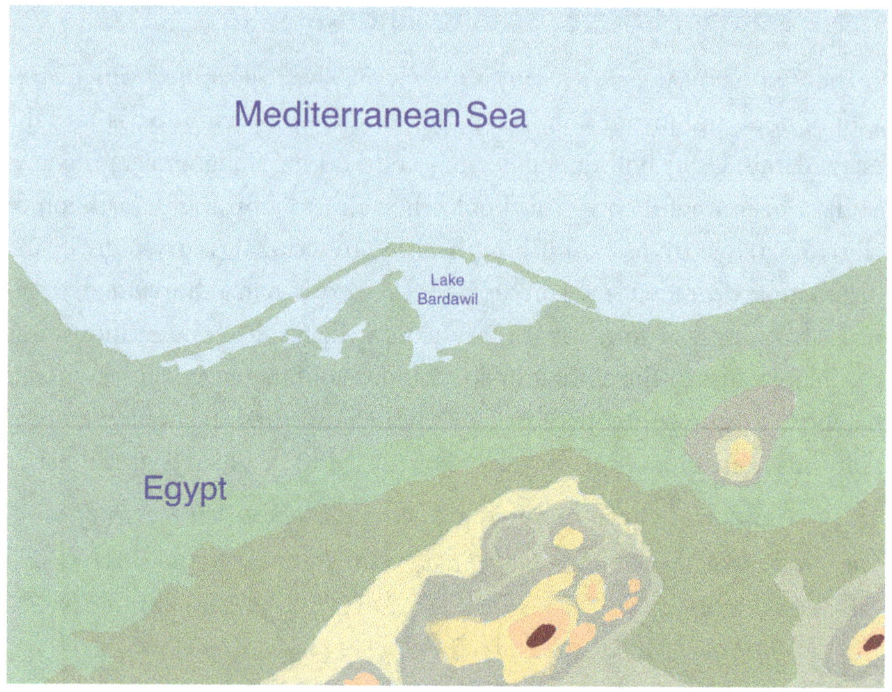

Lake Bardawil is a freshwater lake on the Sinai Peninsula, which has become very degraded through sedimentation coming down the barren slopes of the Sinai desert towards the Meditarranean coastline. The Weather Makers were asked to restore the lake to stimulate economic development.

economy as it used to be a rich source of fish. But over the years it has become severely sedimented with organic matter that eroded from the bare desertified slopes above it. The lake which was once 30-40 meters deep is now barely 1-3 meters deep and has become very salinized.

Ties was going through a transformation in his thinking after meeting Tim Flannery and reading his book *The Weather Makers*. While he was considering the Lake Bardawil project, he met John Liu: that meeting created sparks. John Liu had started his career as a journalist, but his life changed when he was sent by the World Bank in 1995 to do a documentary about a massive restoration project in China, the Loess Plateau Watershed Rehabilitation Project. Being involved with the dramatic restoration of a huge desertified area like the Loess Plateau made him feel that there was no greater contribution to society than restoring the richness of our natural assets. As someone who had been involved in conflict areas around the world as a journalist, John saw the social transformation that was made possible when the people of a region had access to a reasonable livelihood through the sustainable management of local natural resources. He began to see a clear link between land degradation and poverty. People competing for depleted resources and the desperation of struggling for survival led to conflict, social and political problems, and wars. He became more involved as a consultant and spokesperson for restoration in countries around the world. He now sees this work as a better way to keep peace than any other way.

Through his involvement in restoration in over 90 countries, he observed that restoration is a process of reverse engineering, where humans have degraded the land. According to John Liu, the evolutionary process of life developing on land begins with organic matter. The bare geological rocks collect bits of organic matter. Once this organic matter gets wet, microbes begin to break it down. Organic matter determines how much water can be retained. Water is required for microbial action. A bit of wet organic matter collects more water and more organic matter which begins to generate biodiversity. Once the rocks are covered with this organic matter, biodiversity keeps increasing. Birds and insects come along, adding more organic waste matter and further increasing biodiversity. The process of evolutionary succession gets underway, with one set of species creating the biological conditions for the next. Organic matter and biodiversity keep increasing with more and taller plant forms, more animal, insect and bird species.

Human presence has tended to do the opposite of evolution. The longer humans have been around the less biodiversity and the less organic matter there has been. The end result of the process is we get back to bare rocks and sand with no vegetation, no water, no life. In order to regenerate a landscape that has gotten to this point, we start by getting some organic matter to stay around long enough to accumulate a bit of water. After that we can help the process to go a little faster, but basically Nature takes it from there. In one of the restoration projects in which John participated in the Taklamakan Desert in China, workers dug a shallow grid of trenches covering kilometers in the sand and placed straw in the trenches. As soon as the rain came and wet the straw, vegetation began to appear. As soon as there is some vegetation, more plants can grow.

In the Loess Plateau Project 35,000 square kilometers of barren desert was rehabilitated through a three-step process. Give a hat to the hilltop, give a belt to the hills and give shoes to the base. The hat on the hilltop was trees. The belt to the hills was terraces, creating steps along the hillside so that rain could sink in on the upper surface part of the steps instead of rolling straight downhill. The shoes to the base were dams to collect water before it washed away from the area entirely.

Over time the transformation of the Loess plateau has been dramatic. Once barren rocks and sand are now covered with lush vegetation. Insects, birds and animals have returned. Farmers can make a good living growing crops on the fertile terraces. Humidity has returned to the air and water cycles have been restored. Most of all, the story of the Loess Plateau shows that large scale restoration of a very degraded ecosystem is possible.

Ties van der Hoeven recruited John Liu into The Weather Makers to bring his considerable experience to the Lake Bardawil Project. Through discussions with many experts, the group had begun to see that this project could be just the seed of something even bigger. They began to see that restoring Lake Bardawil could be the first step of several interventions which could regreen the Sinai Desert. If they could regreen the Sinai Desert, it could change weather patterns affecting a much larger area, stretching throughout the Mediterranean area over to the Indian Ocean. The presence of vegetation and more humidity in the Sinai Peninsula could, according to climate models, change the flow of winds and alter rainfall patterns in a way that would be of benefit to a very large number of people, countries and ecosystems. They also saw it as a very positive contribution to controlling the environmental

The Eco Oasis designed by the Weather Makers with the collaboration of John Todd is a series of transparent tanks where the biology breeds microorganisms to help restore the biology of degraded water bodies and purify the water. It is housed in a greenhouse which harvests fresh water through condensation.

crisis and all the suffering that is being caused by climate disruptions. Hence the name The Weather Makers. The project Green the Sinai was launched.

The group continues to recruit partners and advisors from around the world to help with the initial stages. The first steps are all about planning and research before any work begins at large scale on the ground. Another pioneer of water restoration was added to the team. And another John, too! John Todd is an American who has spent his career purifying water with what he calls Living Machines. Living Machines are systems which use biology such as plants, microbes and water invertebrates to purify water. Since the 1970s John Todd has worked around the world setting up water purification systems involving a series of transparent above ground tanks filled with contaminated water. The water contains plants and other biological aquatic forms of life which decompose the toxins. After passing through the series of tanks one by one, the water emerges from the system clean enough to pass any environmental standard. John Todd began with living machines for all different types of waste water, from sewage to industrial, and also works on degraded bodies of natural water. With The Weather Makers, the team designed the Eco Oasis, a series of transparent water tanks housed in a greenhouse, which serve as a breeding ground for microorganisms which will help to kickstart ecosystem regeneration by providing the right conditions for a healthy ecological balance. In addition to kickstarting the biological processes that will lead to restoring the aquatic ecosystem, the Eco Oasis also recycles water from condensation inside the greenhouse. Having enough fresh water to start the growth of some vegetation in a very arid environment can be a critical element to begin to reverse the vicious cycle of degradation.

There are five approaches in the plan to regreen the Sinai, and each step is now being fine- tuned and researched. All five steps will happen in parallel once the work begins.

Step one is to restore Lake Bardawil. This will be done by dredging the sediment and deepening the inlets which channel water into the lake. The Eco Oasis will help to rebuild the ecology of the water. With a deeper lake and better water quality, the fish population will return.

Step two is to restore the wetlands around the lake. The organic matter from the lake will be used to revive the ecology of the dehydrated wetlands, and they will come back to life when the deeper, cooled lake and the flow of water from the mountains south of the coast provide enough water to fill them again. Research is being done on salt tolerant plants to filter the water and stabilize the banks.

Step three is to reuse the sediment which is dredged from the lake. Researchers are looking at techniques for separating the materials from the sediment into three categories: organic matter, which will be used to make soil mixes and which will be placed further up the watershed to start the growth of vegetation. The second category is the cohesive matter, such as clay, to be used to bind together structures uphill and to slow, spread and sink the monsoon rains when they come. Thirdly there will be granular material like sand and stones which will also be used for the structures.

Step four is to regreen the desert to the south of Lake Bardawil. This will be done by constructing dykes and dams to capture water uphill, as was done at

Al Baydha. There is also a plan to install fog nets at 700 feet of elevation to capture water from the air. Water is needed to start establishing some hardy vegetation. Once the vegetation has taken a hold, it will be easier to start the succession of plants able to survive. Eventually the plan is to set up natural areas interspersed with areas of agroforestry and managed grazing. It is estimated that it will take 20-40 years to bring the desert to this point of productivity.

Step five is to restore the watershed that feeds the area, beginning with the northern side between Lake Bardawil and the highest peaks in the south of Egypt. This will be done by water harvesting, enhancing evaporation with standing water and enhancing evapotranspiration with strategic planting of vegetation.

The regreening of the Sinai will start with Lake Bardawil. Sediments dredged from the lake will be placed uphill in the watershed to start the growth of plants. Plants and dams will hydrate the watershed, changing weather patterns for the entire region from Northern Africa towards the Indian Ocean

Conclusion

I hope that what I have shared in this book gives you hope and inspiration. Perhaps you have shifted your perspective on global warming and are beginning to see from the perspective of complexity. The cause is not simple and linear, just as Nature and ecosystems are not simple or linear. They are complex, self-organizing systems which balance themselves and create the environment on the planet which enables life, as we know it, to exist. Humans have disrupted this balance, not only by burning fossil fuels but by cutting away at the ecosystems which regulate the carbon cycle and the water cycle.

As John Liu put it, the basis of life is carbon and water. Some organic matter (which is mainly carbon) combined with water enables microorganisms to grow. Once this process gets started it evolves and attracts more life, leading to biodiversity. These three things, carbon, water and biodiversity, are the building blocks of life. As long as humans interact with Nature in a way that reduces biodiversity, soil organic matter and water cycles, we will continue on a path of destruction.

The good news is that if we understand how we created the problem, we can understand how to reverse it. The examples I have shared show that even the most degraded of ecosystems can be brought back to life in a relatively short time. Ten years can turn a dry and barren landscape into a thriving and diverse ecosystem. Nature has remarkable powers of regeneration.

Most of us don't live in deserts, but all of us live in places where the lifestyle of the civilized world is cutting away at the natural mechanisms which make this planet hospitable to life. The way we grow food, the way we manage water, the way we destroy the habitat of other species to make room for our infrastructure is leading us in the direction that creates deserts. As the planet warms, continuing on this destructive pathway will accelerate degeneration.

We have a decade to change this together. I have hope when I see the amazing initiatives of so many people all over the world. There are many ambitious and exciting ecosystem restoration projects. The regenerative agriculture movement and the agroecology movement are leading the way to make our food systems work more like natural systems. There is a groundswell of support for indigenous land stewards and social justice. The work on creating a global accounting system to make natural assets a valued part of the economy and funnel investment into restoring Nature, is a critical piece in supporting large-scale change.

What can you do as an individual? That is a question to reflect on. Everyone has their special situation and talents. There is surely a niche for you. What happens in this next decade is too important for anyone not to get involved.

Here is what I do as a retired, middle class Canadian. I try to raise awareness by talking and writing about regeneration. I volunteer with a non-profit organization Regeneration Canada, to support the spread of regenerative agriculture. I harvest rainwater to water my garden and try not to use any groundwater. I grow organic food. I support indigenous land stewards by attending events they organize and I contribute money to their causes. I give thanks to water and to the earth and try to make my interactions with the earth respectful. I work with my financial advisor to decarbonize my pension investments and to find green investments available to me.

However humble your means and status, there is something you can do. I invite you to reflect and to find your own way to contribute to this movement to regenerate our planet. Whatever your means, wherever you are, you can play a part. If you can't be directly involved in land restoration, you can raise awareness, support businesses which contribute to regeneration, and vote for green policy. You can contribute through art and culture. You can be an example through the way you live your life and the values you demonstrate.

The examples I have shared with you show that by focusing on water cycles and rehydrating the earth, we can have a measurable impact in a short timeframe. In all of the examples, there was a dramatic difference in less than a decade. We still have to reduce greenhouse gas emissions and sequester carbon, but none of these things will turn the climate crisis around within a decade. Restoring water cycles will have a real impact on the climate change we are experiencing now: drought, wildfires, flooding and violent storms. It may take many decades to bring down the greenhouse gases in the atmosphere, but restoring the water cycles may save us by softening the severe impacts of global warming and giving us time to rebalance the atmosphere. The efforts needed to restore water cycles also contribute to sequestering carbon and preserving biodiversity. But if we focus our efforts only on reducing emissions or mechanically pulling carbon out of the atmosphere with technology, we will fail. We cannot continue with business as usual, living in an unsustainable industrialized economy. Unless we restore water cycles and the natural ecosystems which sustain them, we will suffer from continuing climate breakdown. Unless we restructure the values of our society to work collectively in a spirit of respect for the natural systems which sustain life, we will not succeed in traversing this critical transition without great suffering.

Let us join together to make this next decade one of radical social change and ecosystem regeneration. Let us prove that the human presence on this planet can also be a vector for positive change.

What to do with this book once you don't want it anymore?

1. Give it a second life

Extend the life span of this book by giving it to someone or to a charity or by reselling it. A book with little damage can still be useful before being thrown away.

2. Recycle it well

If it has to be thrown away, put it in the recycling bin so that the paper it is made of can be reused. But be careful, it is important to remove recycling disruptors first:

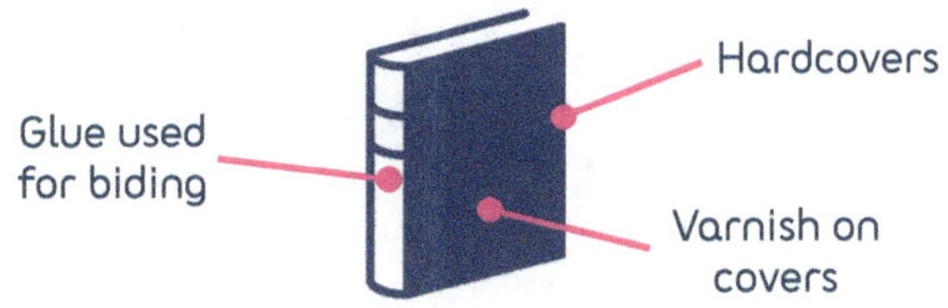

To do this, simply remove its cover and put it in the bin for packagings and all the pages left in the waste paper bin.

Legal deposit: December 2021